MW01131065

Feathered Dinosaurs

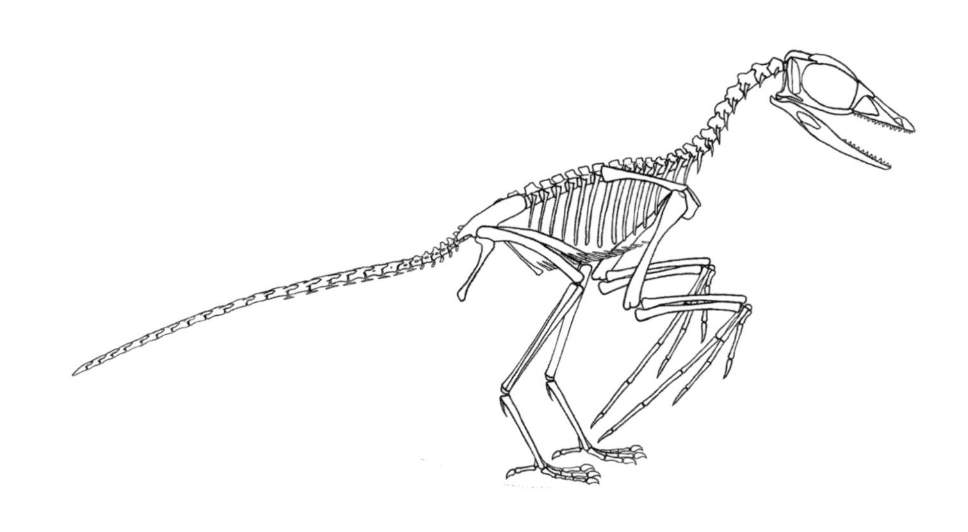

Stephen A. Czerkas
Sylvia J. Czerkas

CONTENTS

**Published by The Dinosaur Museum,
a not-for-profit 501 (c) (3) Public Charity
founded in 1992**

The Dinosaur Museum, 754 South 200 West, Blanding, Utah 84511-3909
http://www.dinosaur-museum.org

ISBN 978-1-932075-03-8
Hardcover
LCCN 2008920857

First printing, February, 2008

Printed in the United States of America on acid-free paper

This book is based on the scientific publication "Feathered Dinosaurs and the Origin of Flight", The Dinosaur Museum Journal, Volume 1, 2002, and the traveling museum exhibitions of "Feathered Dinosaurs" organized by:

The Dinosaur Museum, Blanding, Utah, USA
The Beipiao Fossil Administration Office, Liaoning, China
The Beipiao Shihetun Museum of Paleontology, Liaoning, China
With the cooperation of:
The Institute of Geology, Chinese Academy of Geological Sciences, Beijing, China
The Bureau of Land Resources Management of Chaoyang City, Liaoning Province, China
Beipiao Paleontological Research Center, Liaoning, China

HOSTS OF THE EXHIBITIONS

Feathered Dinosaurs and the Origin of Flight

San Diego Natural History Museum, California
February 7, 2004 - January 2, 2005
Royal Ontario Museum, Toronto, Canada
March 12 - September 5, 2005
Arizona Museum of Natural History, Mesa
March 1 - September 28, 2008
Fresno Metropolitan Museum, California
November 1, 2008 - March 1, 2009

Feathered Dinosaurs Before and After *Archaeopteryx*

Las Vegas Natural History Museum, Nevada
June 2 - October 30, 2005

FEATHERED DINOSAURS

The origin of birds, and how flight evolved, have been great scientific mysteries ever since the ancient bird *Archaeopteryx* was discovered in the 1800's. The relationship of birds and dinosaurs is a major part of this puzzle which has been widely debated and speculated upon. This book and the exhibitions present some of the most extraordinary new fossils from the famous fossil beds of Liaoning, China. With a quality of preservation once thought impossible, the fine lake sediments have retained the most delicate of details. These fossils help solve the mystery of how birds developed feathers and the ability to fly, which reveals a new understanding in how dinosaurs are related to birds.

Mythical dragon or dinosaur?

EARLY HISTORY OF DINOSAURS

The origin of birds and how they evolved their ability to fly has been one of the greatest scientific mysteries. Complicating the mystery is the question of how birds and dinosaurs might be related. Did birds evolve separately from dinosaurs? Or did birds evolve directly from certain types of dinosaurs?

Fossils provide the clues for finding the answers to these questions. However, it can often be difficult to properly interpret the evidence. During recent years though, many remarkable fossils have been found in China which are adding significant details towards understanding the relationship between birds and dinosaurs. The evidence is mounting and has created a tale of discovery with surprising twists and turns.

Throughout history, discoveries of huge fossil bones and footprints probably contributed to myths of giants and dragons. The earliest concepts of these monsters relied more on imagination than a scientific understanding. Perhaps fossil trackways left in stone by huge animals with 3 toes inspired the illustration made in the early 1700's shown at left. If so, then this could be one of the earliest depictions of a dinosaur.

Fossil footprints from the Connecticut Valley, USA.

During the early 1800's, it was generally believed that giant birds had made the fossilized footprints which were being found in the Connecticut Valley. However, when the first fossil bones of dinosaurs were discovered they were thought to be of giant reptiles more like

Early depictions of dinosaurs, circa 1844.

lizards and crocodiles.

When Sir Richard Owen first described dinosaurs in 1841, he distinguished them from other reptiles by their limb posture which he believed to be like that of mammals. This was based on the shape of the femur, the upper bone in the hind legs. Owen observed that the top end of the femur was offset at a distinct right angle. This meant that the leg had to fit into the hip socket in more of a mammal-like fashion than any known reptile. Why Owen did not associate the peculiar legs of dinosaurs with birds may be because he was so influenced by other reptilian characteristics, such as the structure of the teeth

***Iguanodon* by B. Waterhouse Hawkins, circa 1854.**

in *Megalosaurus* and *Iguanodon*. At the time, the thought of birds with teeth was unheard of.

The concept of giant prehistoric birds was known before Owen coined the word Dinosauria, but he did not associate the bird-like fossil footprints as being made by dinosaurs. Instead, he thought that dinosaurs were quadrupedal, walking on all four legs, not just two.

The models of *Megalosaurus* and *Iguanodon* shown above were constructed by Benjamin Waterhouse Hawkins under Sir Richard Owen's direction in the early 1850's. These miniature scale models were then used by Hawkins to make the life-size versions that went on display in 1854 on the grounds of the famed Crystal Palace in England.

***Megalosaurus* by B. Waterhouse Hawkins, circa 1854.**

For the rest of the 18th century, dinosaurs were often portrayed how Owen and Hawkins had envisioned them. However, this original concept of dinosaurs as quadrupeds was changing. The disparity between the smaller size of the front legs compared to the larger back legs led to the speculation that dinosaurs might have been able to rear up in a bipedal fashion as some mammals do. Not too many years after the Crystal Palace dinosaurs went on display, another carnivorous dinosaur was discovered, *Laelaps*, now known as *Dryptosaurus*. It had such small forelimbs that it had to be bipedal. Comparisons with mammals led to the speculation that this dinosaur could jump like a Kangaroo. By the 1870's, Owen's *Iguanodon* had been changed into a two-legged dinosaur as seen in the Victorian school chart below. What is not so readily obvious was what caused this change. It was a new interpretation in which dinosaurs actually stood upright like birds, instead of like mammals.

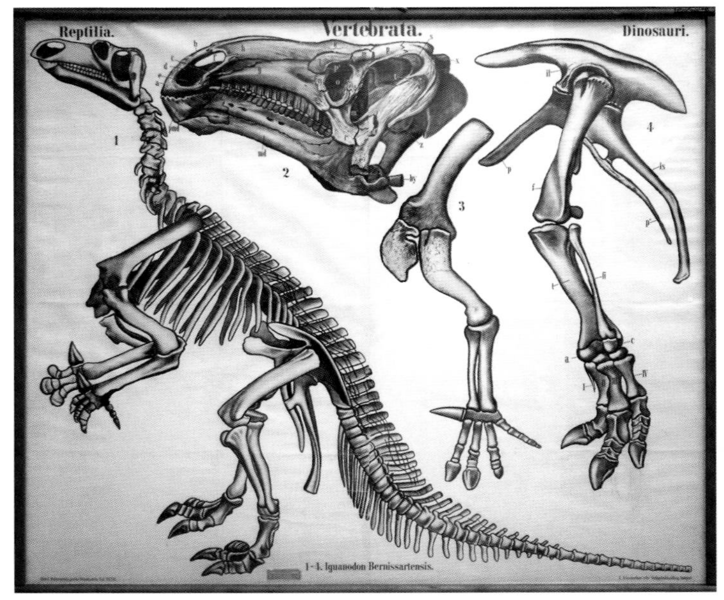

Victorian school chart of *Iguanodon* as a bird-like bipedal dinosaur.

The skeleton of *Compsognathus*.

BIPEDALISM AND BIRDS

By 1868, the fossil of *Compsognathus* led Thomas Huxley to believe that at least some dinosaurs stood upright like birds. Huxley also thought of *Iguanodon* as standing on its two hind legs, but he regarded it as being even more bird-like than *Compsognathus* because it had pelvic bones oriented backwards in an avian fashion.

***Compsognathus* as a scaly bipedal dinosaur.**

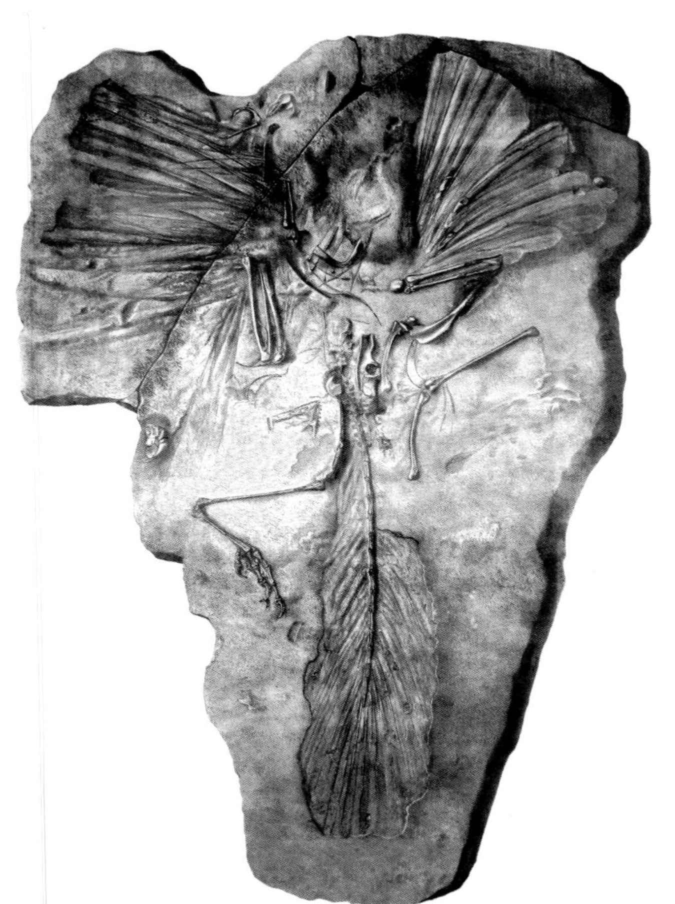

The first fossil skeleton of *Archaeopteryx*.

The first skeleton of *Archaeopteryx* was discovered in 1861, only a couple of years after *Compsognathus* was found. In 1863, Sir Richard Owen identified *Archaeopteryx* as definitely being a bird. The specimen had only a few bones that stood out as being avian. One in particular was a furcula, a bone that is distinctive in birds. Even though it was very small compared to typical birds, Owen recognized the furcula, or "wishbone". The tail of *Archaeopteryx* was most peculiar. It was so long that it resembled a reptile more than a bird. But, the most conclusive evidence for the avian status of *Archaeopteryx* was that it had feathers, actual flight feathers.

Huxley was one of the first proponents of the concept of evolution. And *Archaeopteryx* was a prime example linking birds and reptiles together. But Huxley accepted it as being a bird and not just a feathered reptile. Huxley thought birds evolved from within Dinosauria. His main line of reasoning was based on the observations that various types of dinosaurs were bipedal like birds, and that some, like *Iguanodon,* had pelvic bones which seemed to resemble those of birds. But were these similarities really avian?

***Archaeopteryx* as envisioned by Gerhard Heilmann.**

CONVERGENCE OR HEREDITY?

The problem with Huxley's speculation that birds evolved from within Dinosauria is that similar characteristics might not be due to being passed on by inheritance, but instead can be traits which have evolved independently. The result of two separate lineages that evolved similar traits independently is known as convergence. By having similar functions, or behavior, different lineages can take on characteristics that look remarkably alike. This can make it very difficult to identify how closely related some animals actually are. They may or may not have a distant common ancestor. But even if they did, the development of similar characteristics is not necessarily evidence of a direct relationship. For example, the ostrich, kiwi and kakapo are all flightless birds. They are all birds because they all share a common ancestor which was a bird. However, it would not be correct to think that any one of these flightless birds was the ancestor of the others. This is because they all lost their ability to fly independently of each other. So, when convergence occurs it can be very misleading.

The problems that convergence creates with interpreting how birds might be related to dinosaurs are extremely complex. Did birds inherit specific avian characteristics from some ancestral lineage of dinosaurs? Or did the similarities seen between birds and dinosaurs evolve independently? If birds evolved from dinosaurs, how did that come about? Could the similarities between birds and theropods be due to factors of convergence? If so, what would it take to be able to demonstrate that any resemblances were coincidental and not due to heredity? Did birds evolve directly from a single dinosaurian hierarchy, or family tree? Or could birds have evolved from a separate lineage other than that of dinosaurs?

For most of the past century, it was largely believed that birds and dinosaurs were not directly related and similarities between the two groups were coincidentally brought about by having similar behavior. However, the idea of convergence has fallen out of favor and for over three decades, birds have been thought to be direct descendents of theropod dinosaurs.

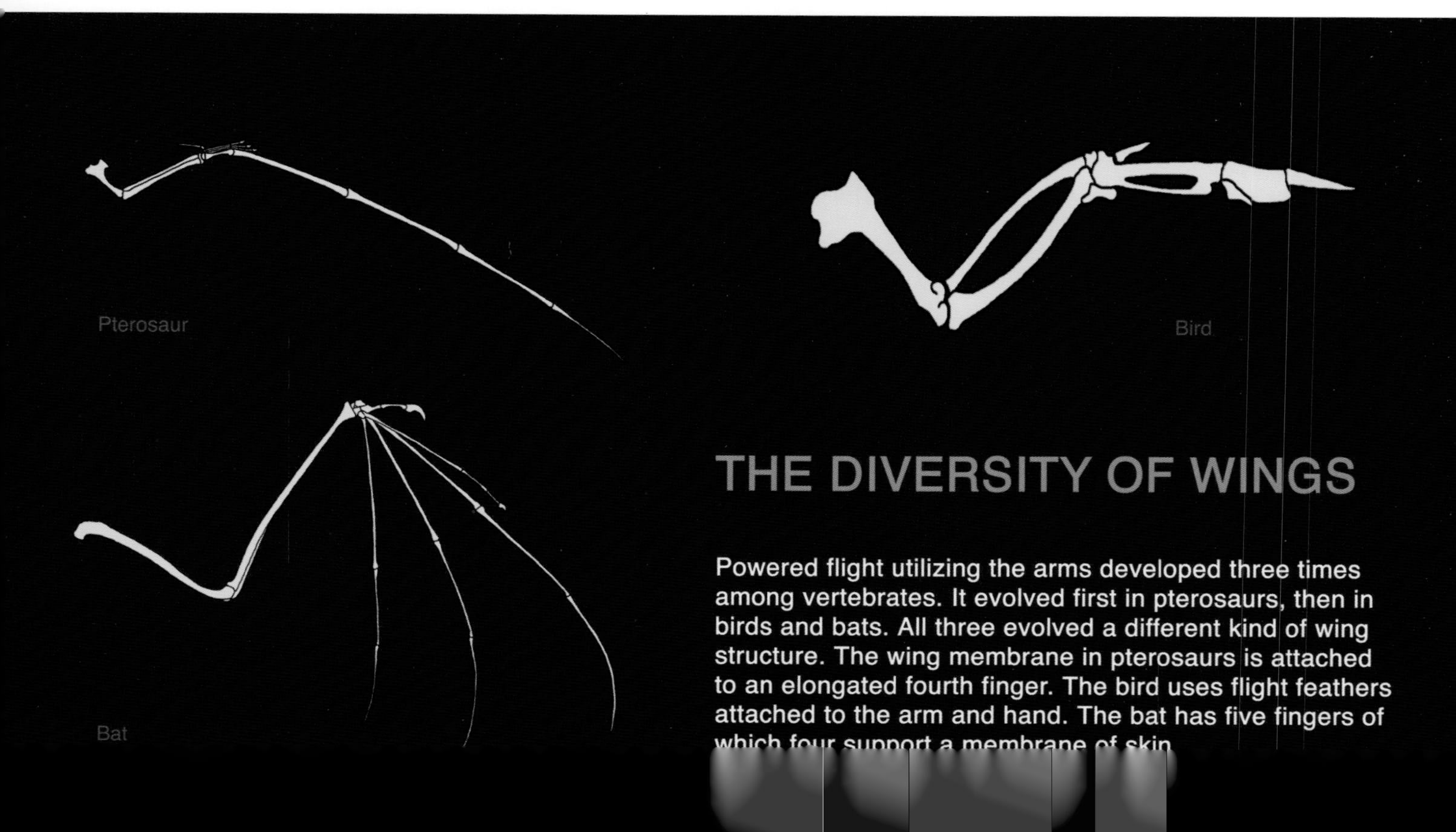

Gliding lizard, Draco

Flying fire bat

FORMS OF FLIGHT

Gliding gecko

Flying stick insect

The ability to fly is a beneficial adaptation which has evolved many times throughout the animal kingdom. Insects were the first to be able to travel through the air. Long before the first dinosaurs appeared, some small arboreal reptiles developed the ability to glide through the air much like Draco, a "flying" lizard of today. Mammals are the most recent vertebrates to evolve the ability to glide, as in squirrels, or to actually fly, as bats do. How birds first evolved the ability to fly is very controversial, but all other flying vertebrates appear to have had arboreal ancestors. There are many examples of arboreal animals that have evolved the ability to glide through the air. Pterosaurs and bats must have had arboreal ancestors that could glide before the ability for powered flight by the flapping of wings was achieved. It would be an extraordinary exception to the rule if birds did not follow the same pattern. Unlike other kinds of animals which evolved the ability to glide or fly, birds are unique in that some have lost their ability to fly.

Gliding squirrel

Gliding frog

Kiwi, a flightless bird

Kakapo, a flightless parrot

FROM THE TREES DOWN, OR FROM THE GROUND UP?

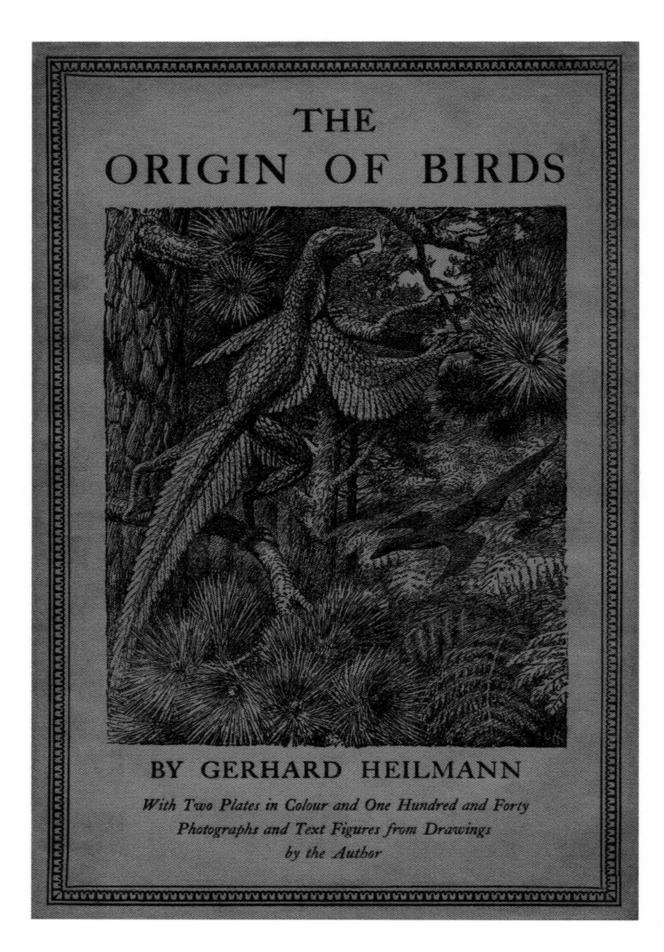
THE ORIGIN OF BIRDS

BY GERHARD HEILMANN

With Two Plates in Colour and One Hundred and Forty Photographs and Text Figures from Drawings by the Author

Above: Heilmann's book created the traditional view of an arboreal ancestor of birds.

For most of the last century, it was thought that birds evolved from ancestors which could climb. It was also believed that the ancestors of birds were not dinosaurs, but were from a more ancient lineage which predated dinosaurs. By being able to climb, these "arboreal", or tree-dwelling ancestors of birds went through different stages from leaping, which led to gliding, and then eventually to being able to fly. This process had the advantage of using gravity to help in propelling the animals through the air. What was lacking to support this concept of how birds evolved was any fossil evidence to show that the arboreal ancestors had existed.

In the mid-1960's, fossils of a small dinosaur, called *Deinonychus*, were discovered. It was believed to be a dinosaur, and not a bird. But *Deinonychus* was thought to be somehow related to birds because its skeleton appeared to have avian characteristics. However, since *Deinonychus* could not fly, the similarities it had with birds were believed to represent ancestral stages of development before flight had been achieved. What this meant was that some dinosaurs, like *Deinonychus*, had the potential to be ancestors of birds even though they were "cursorial", or ground-dwelling animals.

Left: Even though *Deinonychus* was thought of as a scaly dinosaur, it led towards the concept that birds evolved from the "ground up". Right: The hand of *Deinonychus* had a bone in the wrist which allowed a peculiar range of motion like that in the wing of a bird.

LIAONING: STORIES IN STONE

Fossils represent evidence of life from the prehistoric past. These relics in stone provide tangible clues which make it possible to understand how life evolved. Some kinds of fossils have been preserved in abundance, while others can be extremely rare. Among the most scarce are fossils of birds, especially those that lived during the Mesozoic, the age of dinosaurs. However, in the northwest province of China called Liaoning, there have been an unprecedented number of fossils discovered in recent years, including those of ancient birds.

These fossils are often astounding in how well they are preserved in the fine-grained rock sediments which are composed of volcanic ash. This suggests that volcanos not only killed these prehistoric animals, but also preserved them so remarkably through the ages.

The fossils of Liaoning are not only phenomenal in the vast quantity they are often found in, but also for their remarkable quality of preservation which is among the finest anywhere in the world. What makes the fossils of Liaoning stand out the most is the stories that they tell and the evidence they provide for the scientific understanding of prehistoric life during the age of dinosaurs. Looking at these fossils is like looking back in time some 125 million years. The closer they are studied, the better the details of a world lost in time can come into focus, putting the story of prehistoric life into perspective with that of our own time.

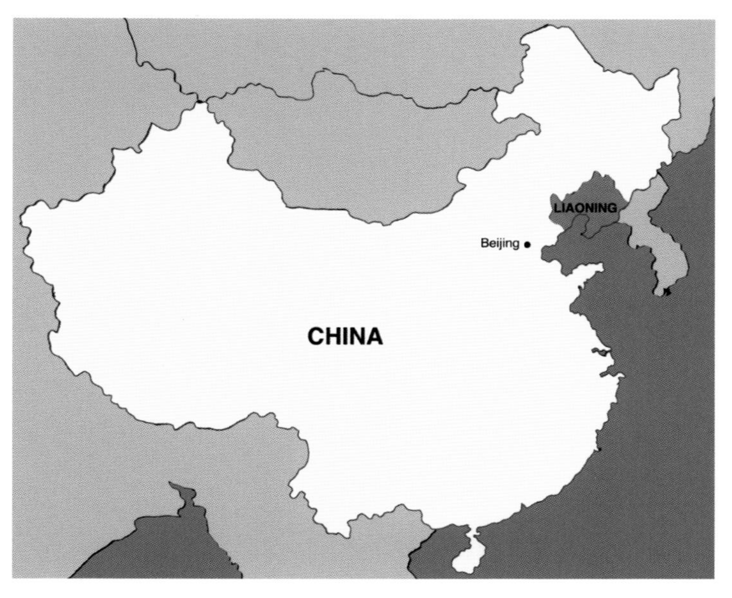

Above: The map of China indicating where Liaoning is located. Bottom left: Two fish, *Lycoptera*, preserved with a branch of a conifer tree, *Pityocladus*. Below: A chart depicting where the fossils of Liaoning are from in time.

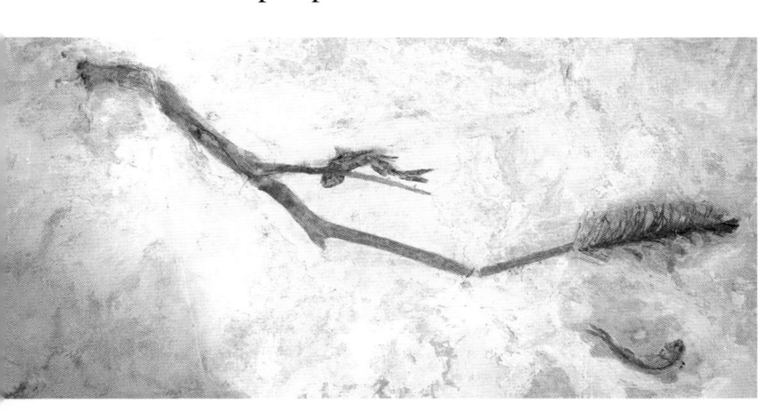

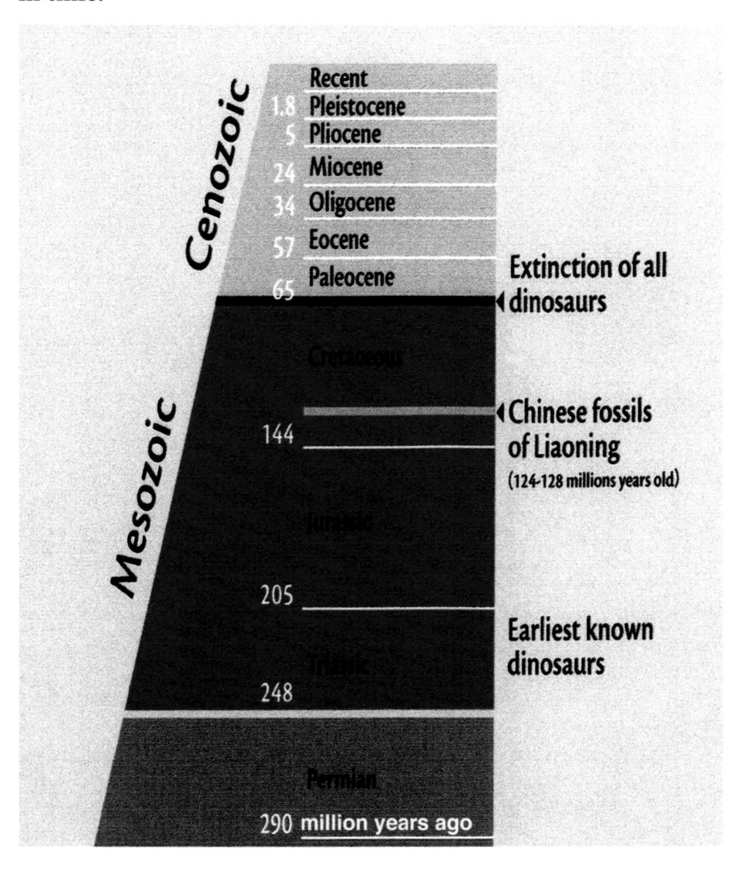

A STRANGELY FAMILIAR ECOSYSTEM

Many of the fossils from Liaoning appear remarkably reminiscent to forms of life still living today. What is amazing is that some plants, insects, amphibians and fish appear to be largely unchanged from their prehistoric counterparts. This reveals one of the greatest mysteries of the great extinction which killed off the dinosaurs. The question is: Why did so many varied kinds of life continue to survive while the dinosaurs did not? Finding the answer as to what possibly killed off the dinosaurs is only part of the problem. Explaining how their extinction could occur while letting other kinds of life survive, basically unchanged, is an even greater issue that is yet to be resolved.

Top: a dragonfly with a wingspan of about seven inches. Left: A spider largely magnified.

Todites was a fern that would have been an abundant ground cover.

A salamander-like amphibian, *Chunerpeton,* from the Middle Jurassic.

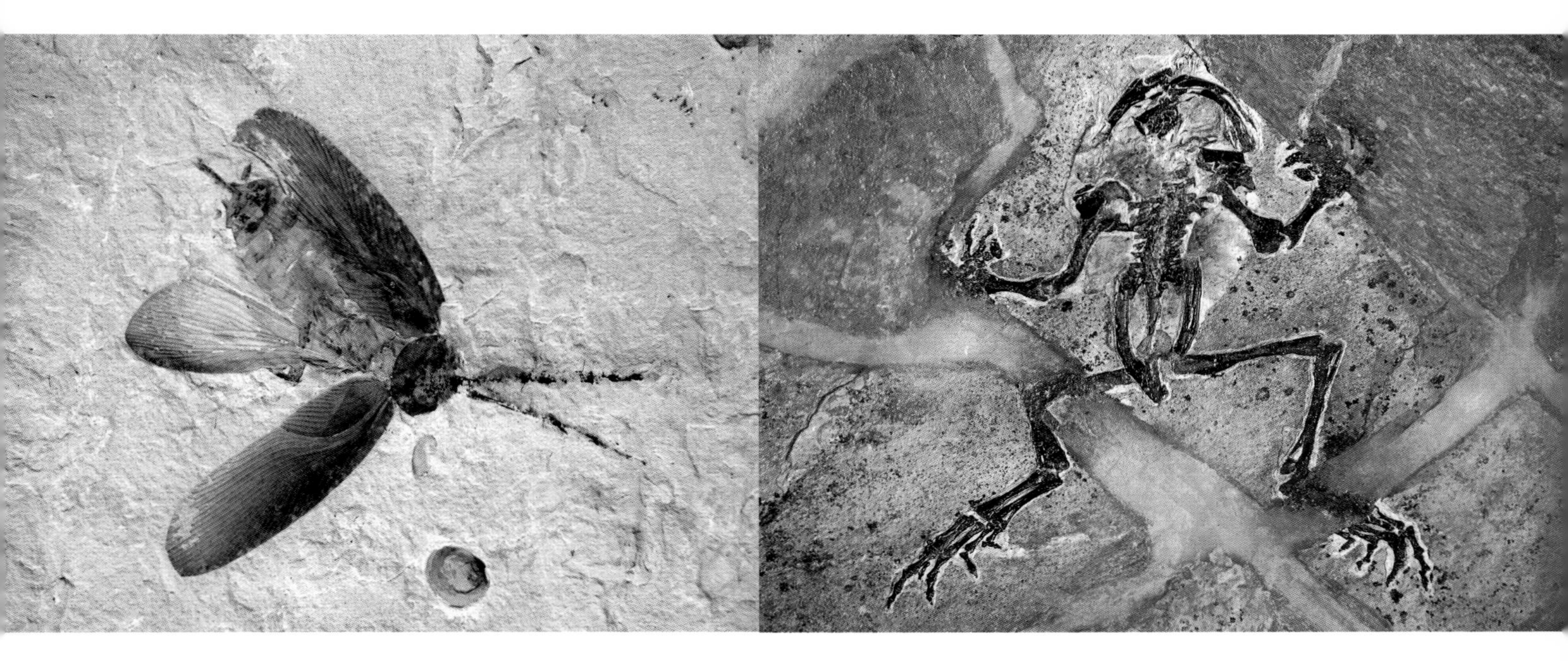

Phipidoblattina, a flying insect resembling a cockroach, from the Middle Jurassic.

Above: A Lower Cretaceous frog, *Callobatrachus*. Below: *Protopsephurus*, a sturgeon and small fish, *Lycoptera*.

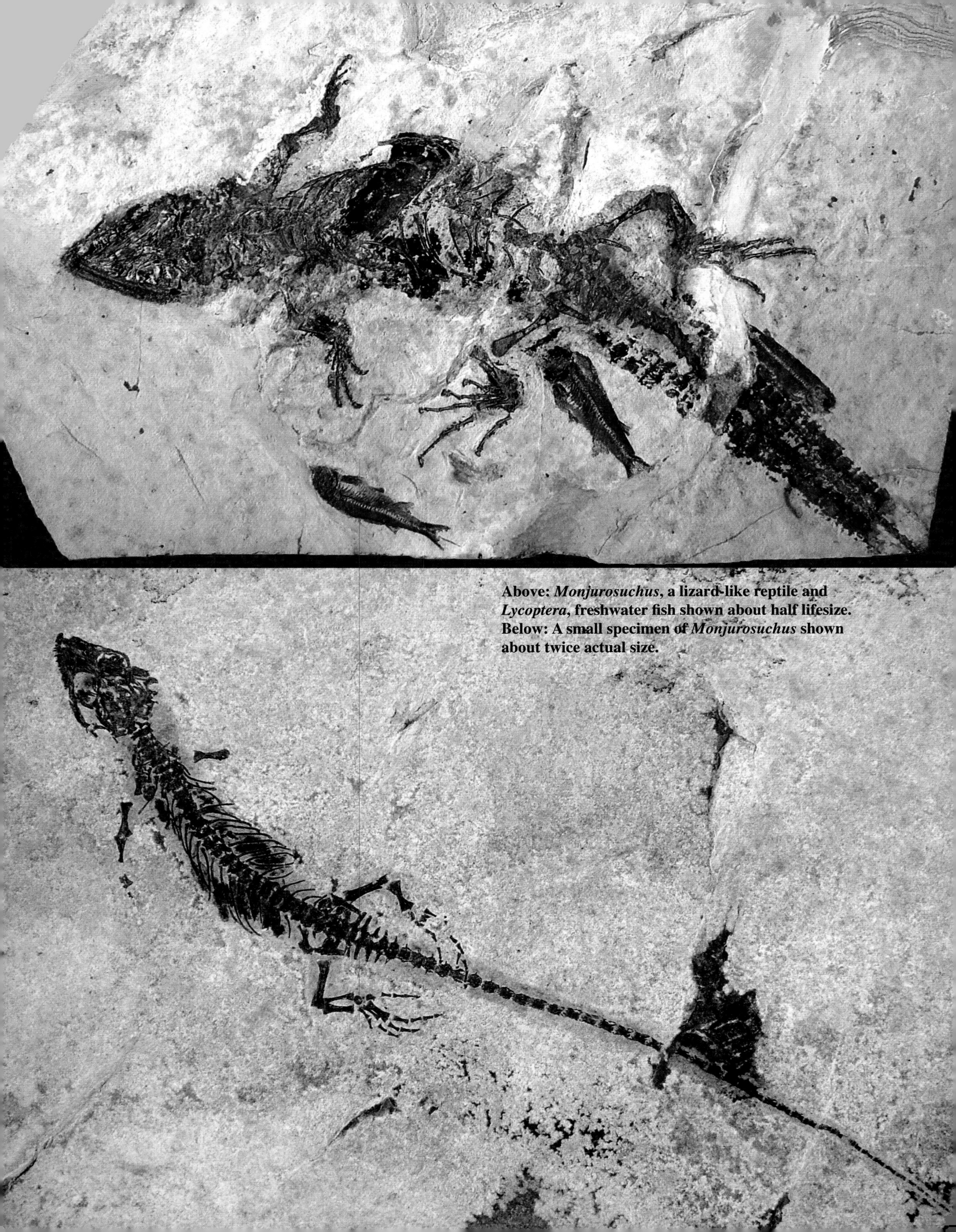

Above: ***Monjurosuchus*****, a lizard-like reptile and** ***Lycoptera*****, freshwater fish shown about half lifesize.**
Below: A small specimen of ***Monjurosuchus*** **shown about twice actual size.**

A diversity of reptiles is well represented by the fossils of Liaoning. In addition to the skeletal remains, it is not uncommon for skin impressions to be preserved. Various types of lizard-like reptiles have been found including some that were well adapted to a semi-aquatic lifestyle. Turtles are abundant as well. Small dinosaurs like *Jeholosaurus* and *Psittacosaurus* may not have appeared too different from other reptiles that they co-existed with, but they were. Both represent early ancestral forms of ornithischian dinosaurs. The highly derived descendants related to *Psittacosaurus* evolved into horned dinosaurs like *Triceratops*.

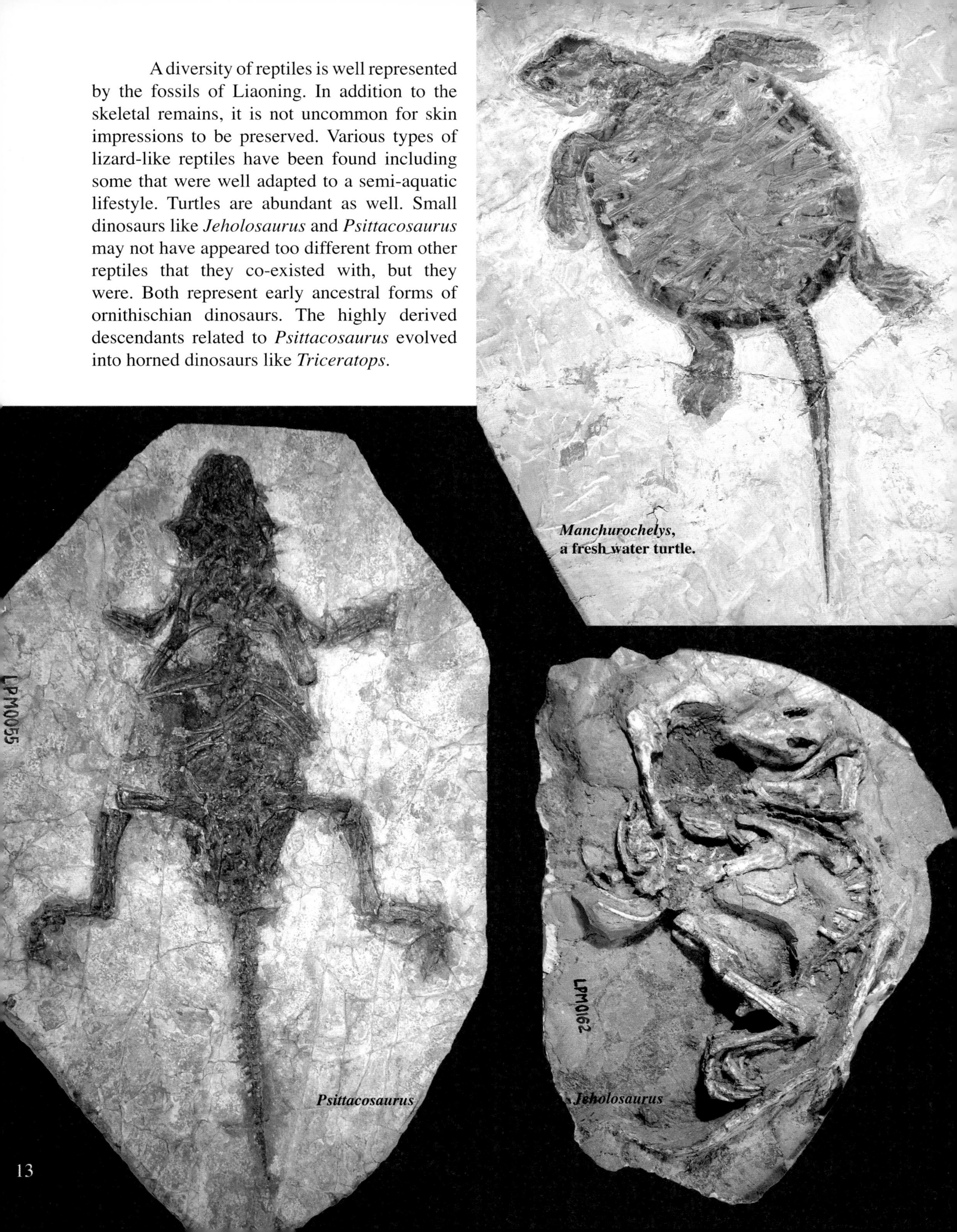

***Manchurochelys*, a fresh water turtle.**

Psittacosaurus

Jeholosaurus

MESOZOIC MAMMALS

The skeletal remains of at least eleven animals are preserved together in what may have been their nest or burrow. They appear to represent two kinds of mammals. Most are relatively primitive in having ribs that extend all the way back to the pelvis. But there is at least one that represents a more modern type which no longer had the ribs of the lower back. Why two kinds of animals were preserved together is a mystery. However, this is the largest group of Mesozoic mammals ever found, and it suggests that gregarious behavior existed in these rare early animals.

Fossils of mammals are extremely rare even in the quarries of Liaoning. However, the few discoveries which have been made have revealed that a much broader diversity of mammals existed by this time of 125 million years ago. Some of the mammals have been much larger than what have been previously discovered in the Mesozoic. Approaching the size of a cat or small dog, one carnivorous form has been found with stomach contents indicating that it fed upon small psittacosaurs. Even more amazing are the discoveries showing the broad diversity of lifestyles that prehistoric mammals had already achieved. One had adapted itself to an aquatic way of life to the extent that it looked like a cross between an otter and a beaver. And to the other extreme, an arboreal form had already evolved the webbing of skin along the sides of the body which made it possible for it to glide through the air much like a flying squirrel.

The ability to fly is a remarkable adaptation for any kind of animal. And yet, by the beginning of the Cretaceous, the sky was already alive with gliding reptiles, pterosaurs, birds, and even mammals.

FLYING REPTILES

Pterosaurs are prehistoric reptiles but they are not dinosaurs. These flying reptiles represent their own separate lineage. They originated from ancestors that were older than what led to the Dinosauria. The fossils of these mysterious ancestors of pterosaurs have yet to be discovered, but they are believed to stem from the Archosauria. Dinosauria must have evolved from archosaurian ancestors as well. While this might suggest the possibility of both having a common ancestry, just exactly how close pterosaurs and dinosaurs are related remains a matter of speculation.

The pterosaurs from Liaoning represent a diversity which is similar to what is found elsewhere, especially in Europe. The pterosaurs, *Eosipterus*, closely resemble the better known *Pterodactylus* which have been found in the limestone quarries of Solnhofen, Bavaria. *Jeholopterus* is a very peculiar pterosaur in having a rounded skull. It also has European counterparts. The specimens of *Jeholopterus* from Liaoning reveal that it was covered in what appears to resemble "hair". But was it really?

Jeholopterus*, a "frog-mouthed" pterosaur.

A life restoration of *Jeholopterus*.

The neck, body, legs and wing of *Eosipterus*.

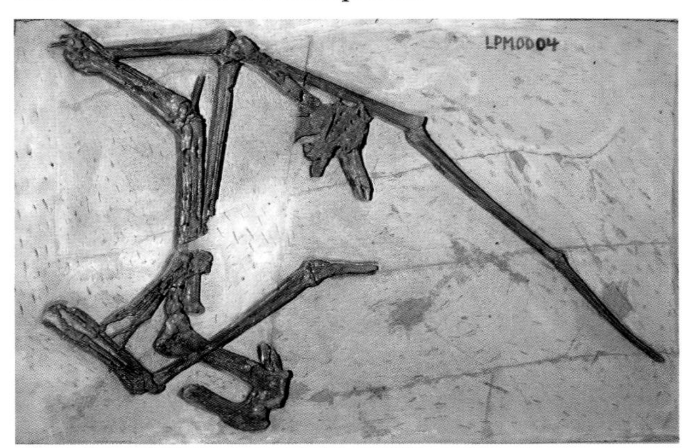

A complete wing, pelvis and much of both hind legs to a specimen of *Eosipterus*.

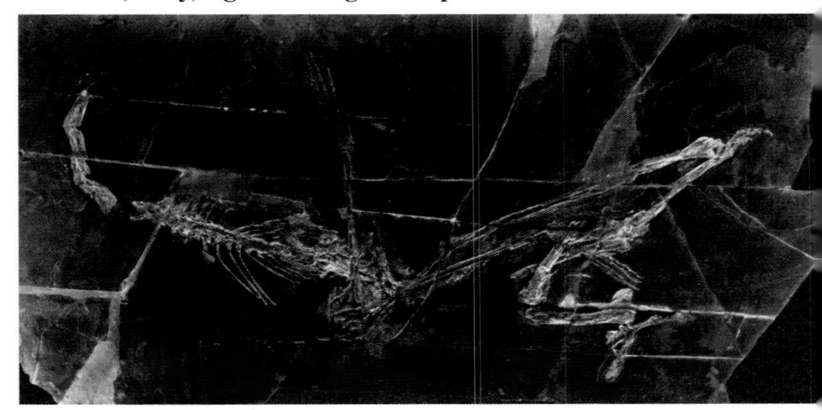

The same specimen of *Eosipterus* seen directly above, but in ultraviolet light.

FEATHERS, BUT NOT THOSE OF BIRDS

Feathers have always been the most diagnostic characteristic of birds. Simply put, if an animal was covered by feathers, then it was undoubtedly a bird. But fossils from Liaoning are making this distinction less reliable, because some of the animals that appear to have been covered by feathers are clearly not birds.

The first of these problematic feathered animals was *Sinosauropteryx*. It was a relatively small theropod, or carnivorous dinosaur. Two specimens had been discovered and both were preserved with what appeared to be short hair-like strands. Because *Sinosauropteryx* is a theropod, a logical implication would be that the strands were primitive feathers and that *Sinosauropteryx* was a feathered dinosaur representing an ancestral condition before birds had evolved the ability to fly.

Sinosauropteryx appeared to be an ideal example for supporting the "ground up" concept of birds having evolved from theropods. Here was a dinosaur without any indications that it or its ancestors ever had any flying ability. So, it was not a bird. But by its having what looked like primitive feathers, there was an obvious implication that *Sinosauropteryx* was distantly related to birds. The real question though is how are they related? Did dinosaurs like *Sinosauropteryx* continue to evolve into becoming actual birds? Or were these kinds of dinosaurs strictly terrestrial and not evolving towards the development of flight?

Complicating the issue is another specimen of a theropod called *Juravenator*. It is also small in size and appears related to *Sinosauropteryx*, but it has no sign of feathers. Instead, it appears that *Juravenator* was covered only by scales.

***Sinosauropteryx*, a feathered dinosaur.**

Above: The fossil of *Pterorhynchus* in UV light.
Below: *Pterorhynchus* as in life.

Many fossils of pterosaurs, the flying reptiles, have been known to have "hair-like" strands or fibers covering their bodies. What these filaments actually are has been unresolved. There is no justification in regarding them as real hair as that is a mammalian trait. However, fossils from Liaoning are so well preserved that answers to this mystery may finally be at hand. The fossil of *Pterorhynchus* reveals that it had a headcrest which was made of keratin, a rigid horn-like material, instead of bone. Also, there were fibers all over the body which look similar to down-like feathers. Since the body was covered, the fibers must have contributed to insulating the body, but a large ornamental beard shows that the strands were also used for ornamentation. The structure of the fibers are strikingly similar to down feathers in young birds. This implies that either birds and pterosaurs inherited feathers from a distant common ancestor, or feathers evolved more than once. Either way, this means that even if the dinosaur, *Sinosauropteryx* was feathered, it still may not have been an ancestor of birds.

DEVELOPMENTAL STAGES OF FEATHERS

A hypothetical reconstruction of the different stages of development in the evolution of a feather can be made by continually simplifying the structure of a modern feather. The most advanced feather is a flight feather that has a hollow root, called the calamus, which extends into a long rachis, or shaft, complete with asymmetrical vanes made up of barbs, barbules, and hooklets. Stage 2 is what we now know existed on pterosaurs. The simple structure of their proto-feathers consisted of only branched barbs extending from a calamus. This means that feathers either evolved independently in different lineages, or birds and pterosaurs share a distant common ancestor.

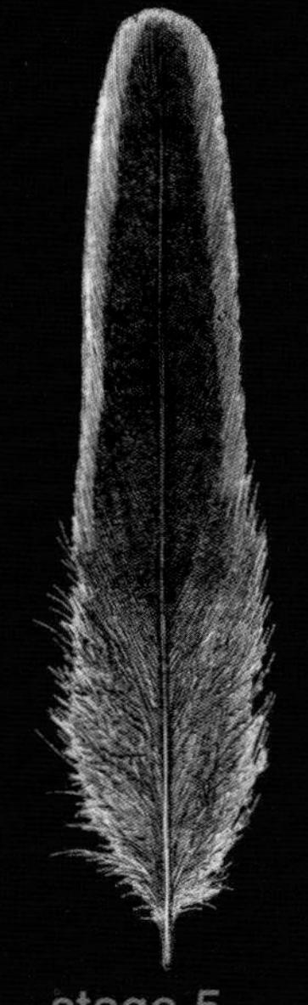

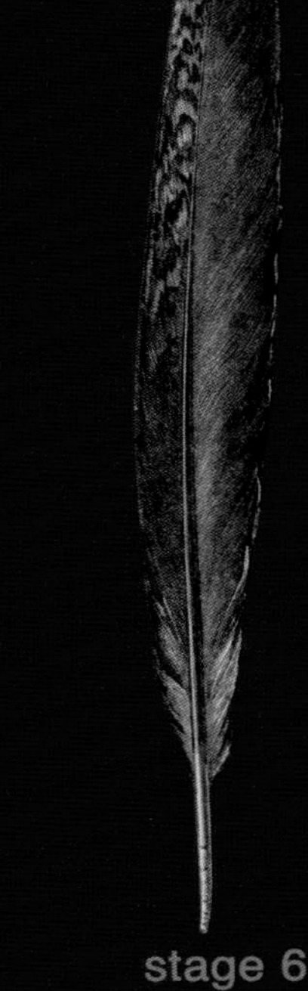

stage 1
calamus and bristle

stage 2
calamus with barbs

stage 3
down feather with barbules

stage 4
plumulaceous vanes and rachis

stage 5
symmetrical pennaceous vanes

stage 6
asymmetrical pennaceous vanes

The mainslab and counterslab fossils of a symmetrical pennaceous feather.

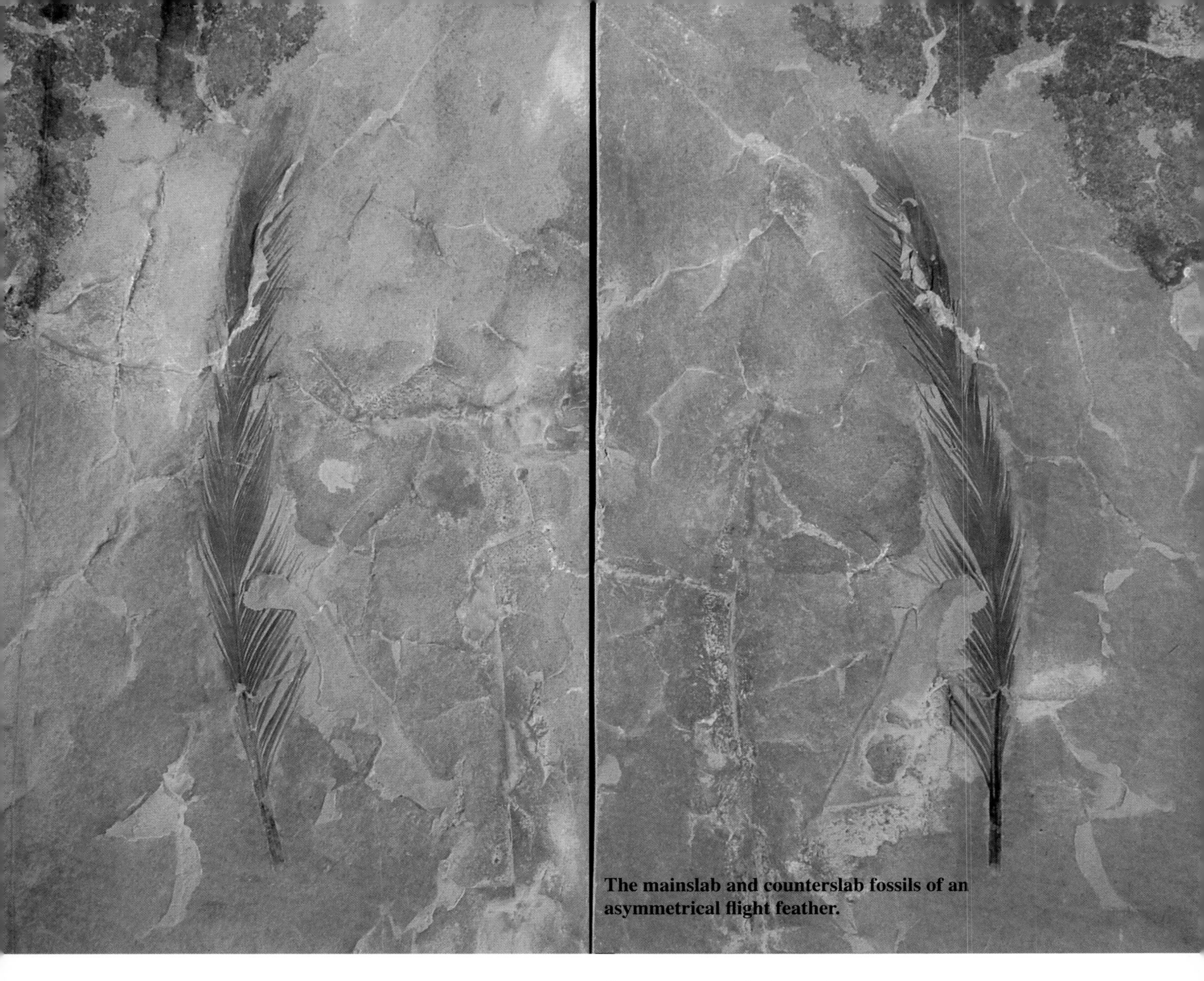

The mainslab and counterslab fossils of an asymmetrical flight feather.

REAL FEATHERS, REAL PROBLEMS

Because of the fine texture of the volcanic ash that makes up much of the sedimentary rock layers which contain the fossils of Liaoning, it is possible that even the most delicate details of a feather can be preserved. Isolated feathers are unmistakable and yet when what looked like feathers started to be discovered on what everyone thought of as dinosaurs, there was an uproar of protests and denials. There was also an enthusiastic claim that here was the ultimate proof that birds had evolved from dinosaurs.

Unlike *Sinosauropteryx* which bears very little resemblance to birds, the dinosaurs that were thought to be the closest ancestral form next to birds are known as dromaeosaurs. They are perhaps better known as the "raptors" that have been popularized in motion pictures. *Deinonychus* was the dinosaur that provided much of the original basis for thinking that birds evolved from similar kinds of dinosaurs. It is a dromaeosaur. The first few fossils of dromaeosaurs found in Liaoning were very controversial because the preservation of their feathers was very convincing to some but not to others. One thing that was universally agreed upon was that dromaeosaurs were not birds. A small dromaeosaur, the size of a bird, was found with a halo of feathery impressions, but no sign of flight feathers on the arms. It looked like a feathered dinosaur, and it was claimed that it was definitely "not a bird". But, this was a serious problem that came with a real surprise.

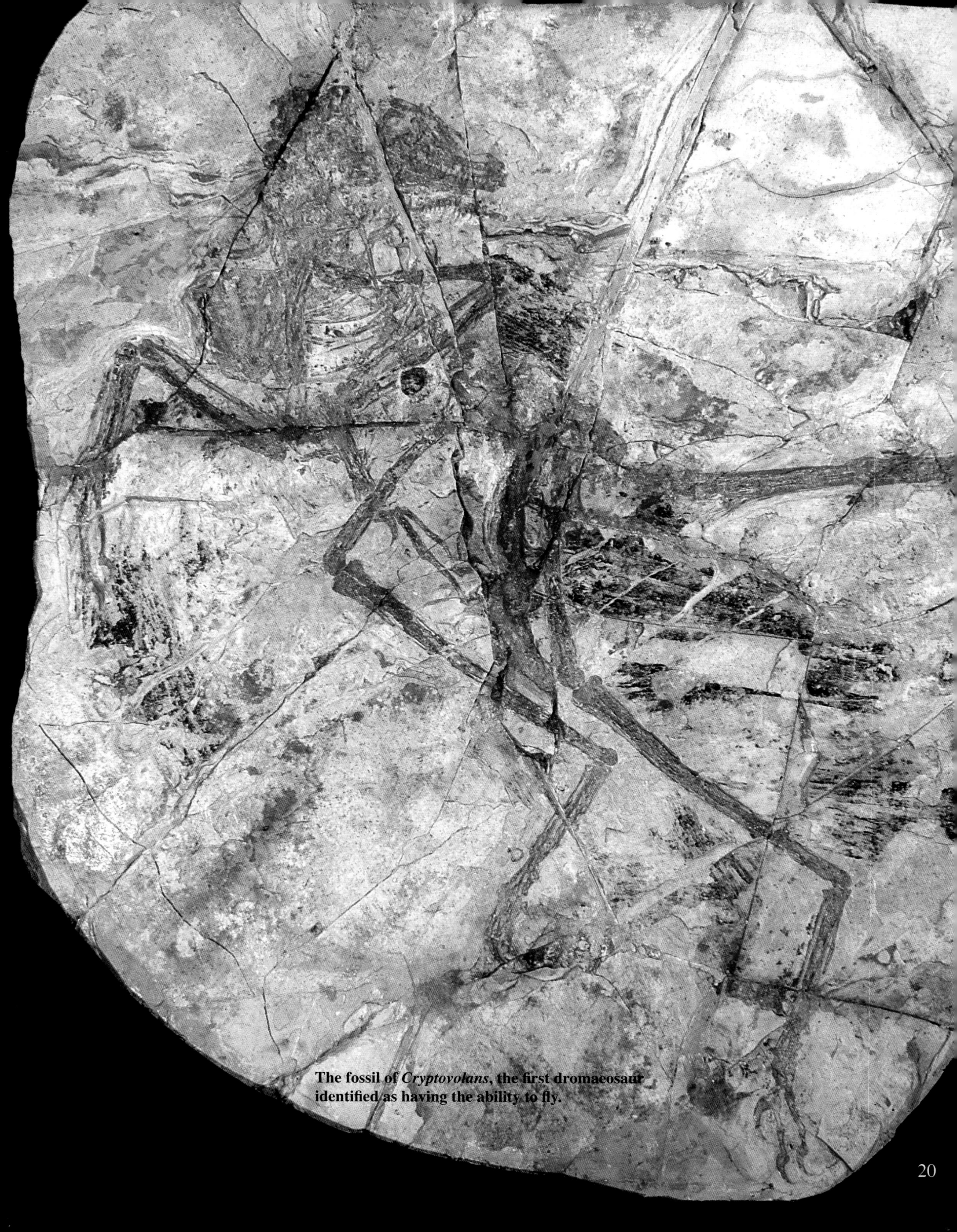

The fossil of *Cryptovolans*, the first dromaeosaur identified as having the ability to fly.

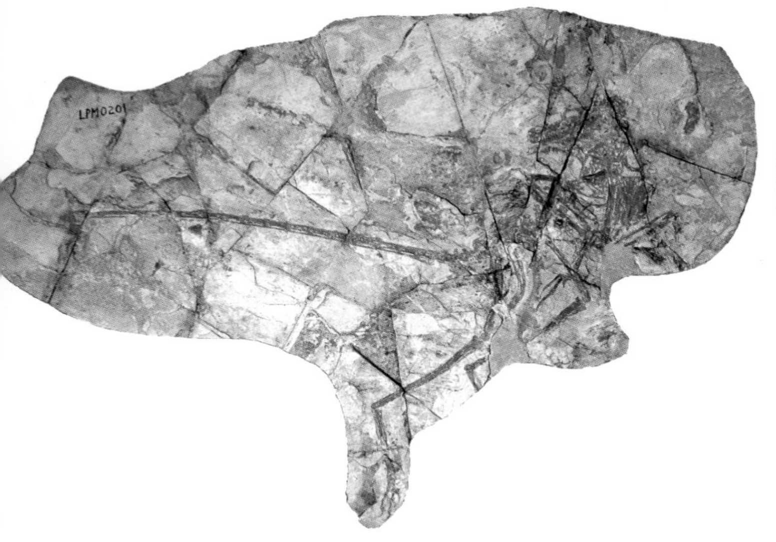

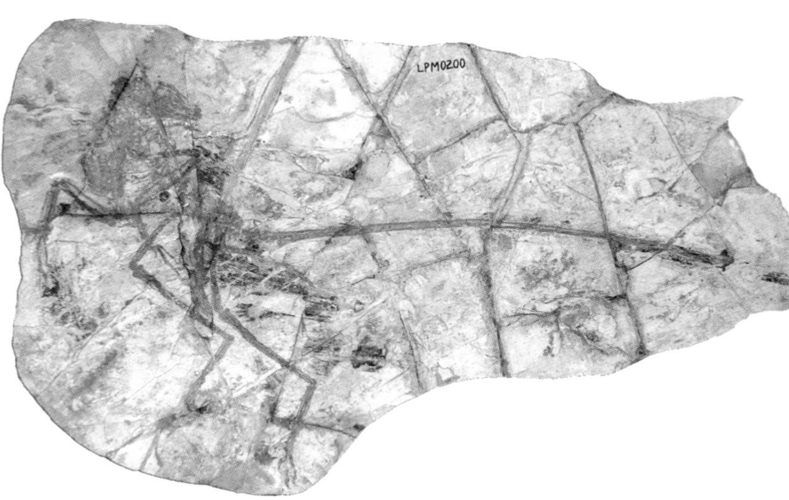

IT'S A BIRD!

Top left: The counterslab of the fossil to *Cryptovolans*. Top right: The fossil mainslab of *Cryptovolans*.

Dromaeosaurs were thought to be the closest type of dinosaur related to birds. They were definitely not supposed to be actual birds, but just a related ancestral form that represented what the pre-avian status of birds was like. The distinction that disqualified dromaeosaurs from being recognized as real birds was that they supposedly did not have any ability to fly. This interpretation was the fundamental basis which helped create the concept that "birds are dinosaurs". However, the specimen on these pages was named *Cryptovolans* which means "hidden flyer". This is because it was the first dromaeosaur preserved well enough to reveal that it had flight feathers which made its arms actual wings. This discovery changes the entire premise of how birds evolved from theropods, because it showed that dromaeosaurs were not dinosaurs that predated birds. They already were birds.

Above: Detail of the teeth and skull in the counterslab of *Cryptovolans*. Below: Detail of asymmetrical feathers on the mainslab of *Cryptovolans*.

After *Cryptovolans* was described, additional specimens of similar dromaeosaurs verified that they were equipped with wings. The flight feathers were asymmetrical and indistinguishable from birds of today that can fly. Since dromaeosaurs were definitely not supposed to have wings or any ability to fly, this was a startling revelation which contradicted the popular belief that the ancestors of birds were dinosaurs. What it meant was that the similarities dromaeosaurs had with birds were not because they were dinosaurs evolving towards becoming birds. That was a mistake. But what undermined drawing attention to this mistake was that dromaeosaurs not only had a regular set of wings, they also had a second set of flight feathers on their hind legs. This extraordinary development seemed to represent a gliding stage in the origins of how birds evolved the ability to fly. This led to the claims that the ancestors of birds were "four-winged dinosaurs". What was lost in the excitement was that dromaeosaurs were not supposed to have even two wings, let alone four.

Parts of the skeletons in these fossils of dromaeosaurs were pulled apart after death, but impressions of feathers can still be associated with the wings, legs and tail.

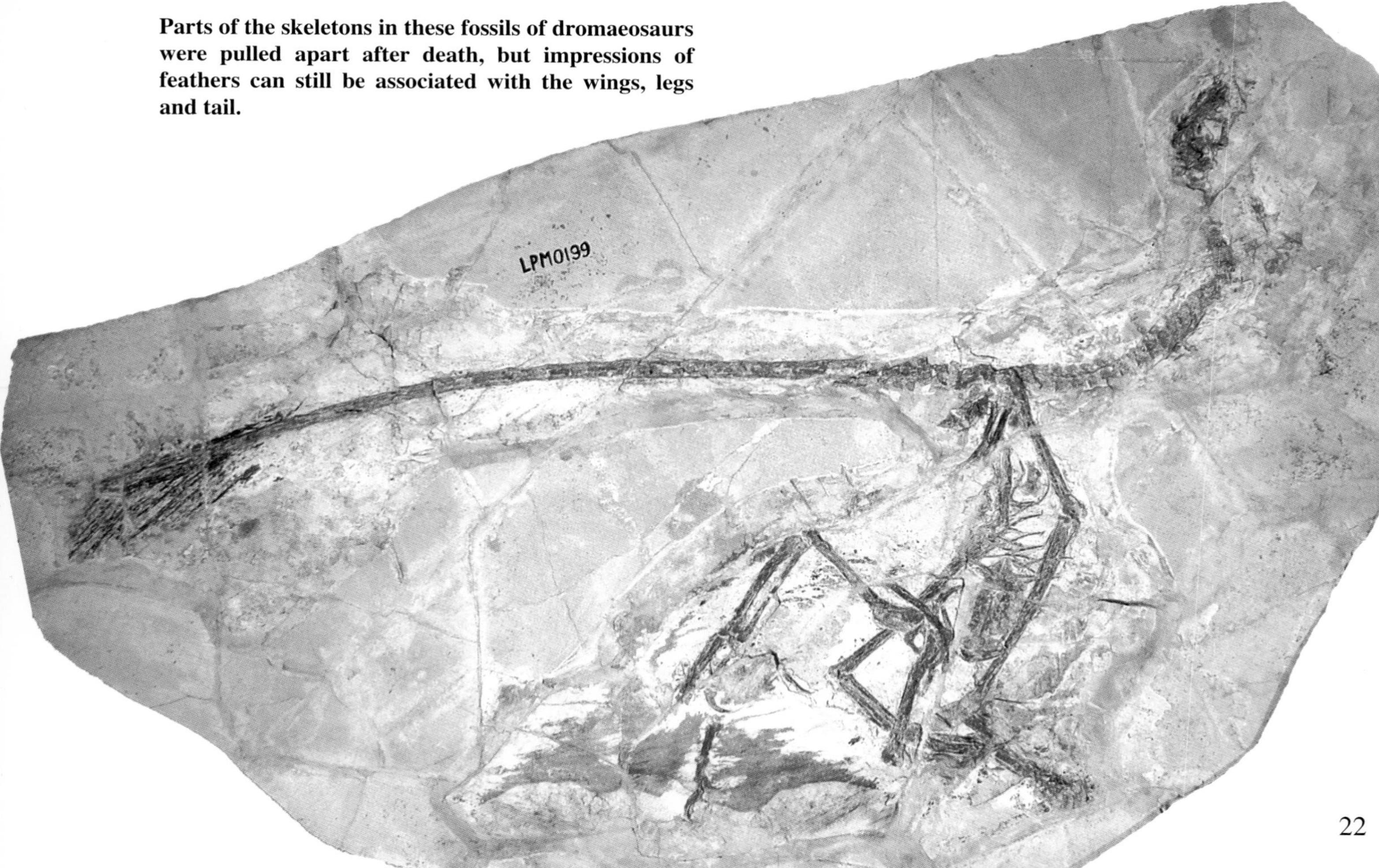

Above: **A life restoration of *Cryptovolans*, a "four-winged" dromaeosaur flying through the air.**

The idea that dromaeosaurs represented the kind of dinosaurs which predated birds was based on the belief that they were not birds. The discovery that some dromaeosaurs could fly revealed that this was incorrect.

The diagram in the upper part of the chart on the right represents the kind of relationship in which bird-like dinosaurs would be ancestral to Aves, the classification which includes all actual birds. As you can see, birds appear to come after dromaeosaurs in this interpretation.

The lower diagram represents the new evidence based on the fact that dromaeosaurs were actually birds because they already had at least some ability to fly. The significant difference is that dromaeosaurs now qualify as being real birds within the classification Aves. And because dromaeosaurs are really birds, this discredits previous interpretations of how birds might have evolved from theropods like dromaeosaurs. Not only are dromaeosaurs birds, but they have a greater development in their flight anatomy which makes them even more avian than *Archaeopteryx*.

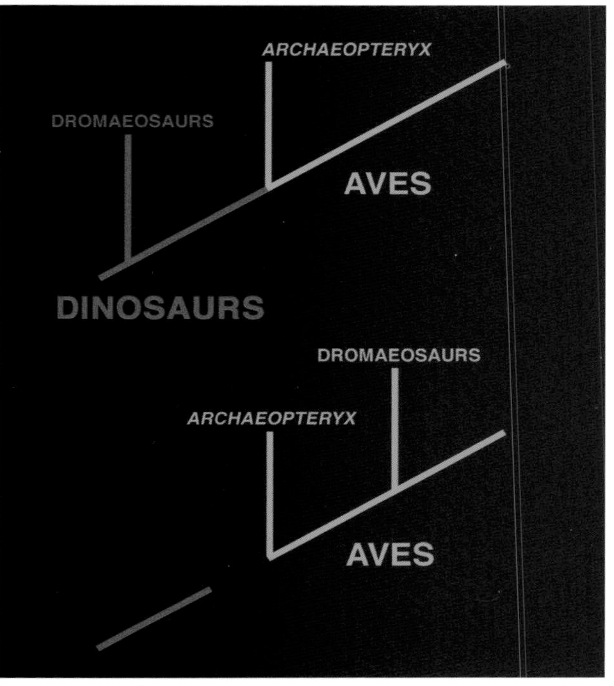

The upper diagram represents dromaeosaurs as bird-like dinosaurs leading towards the evolution of birds. It is incorrect. The lower diagram accurately places dromaeosaurs within Aves and breaks the connection with dinosaurs as being the ancestors of birds.

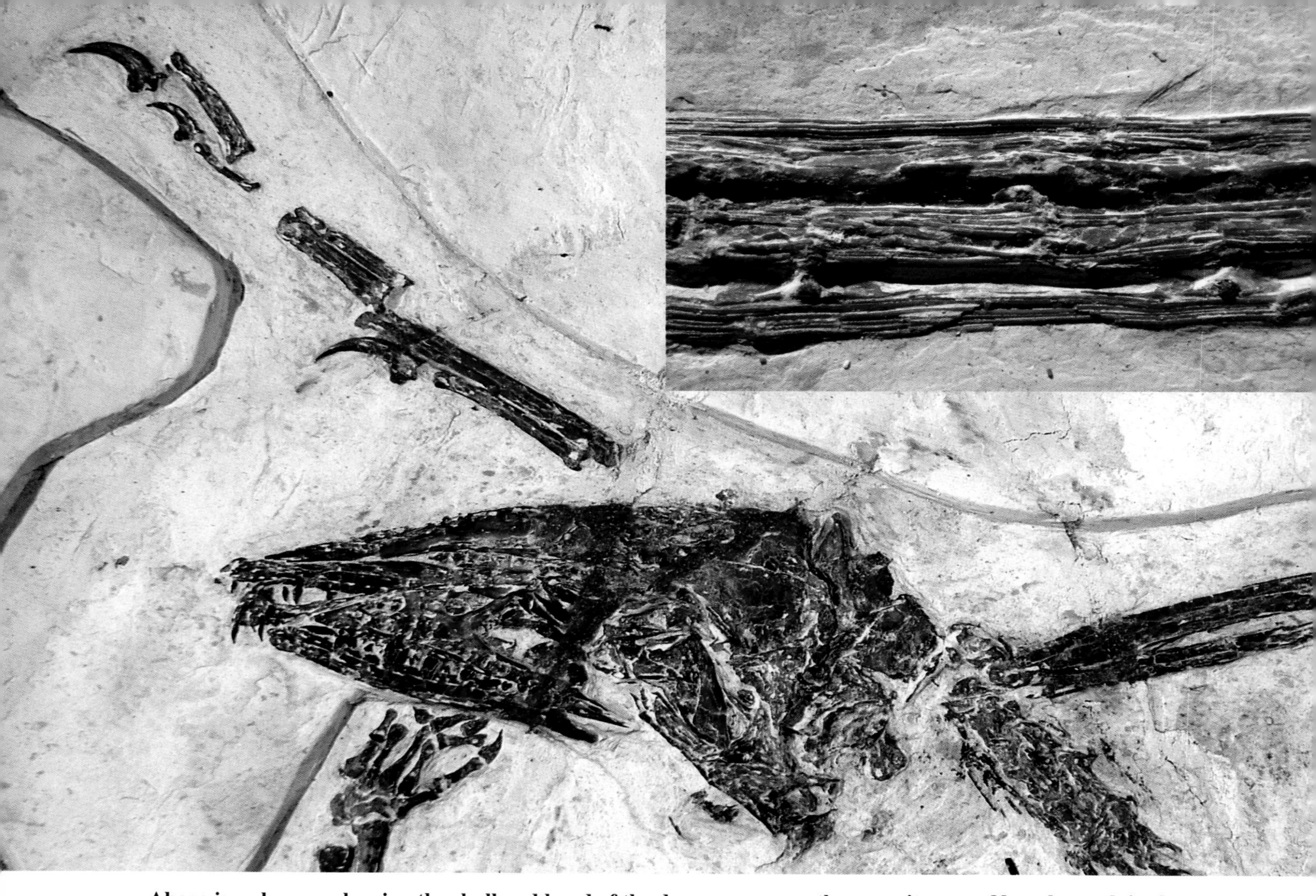

Above is a close-up showing the skull and hand of the dromaeosaur on the opposite page. Note the teeth in the jaws and the claws on three fingers. The last bone and claw of the second finger are located just above the same bones of the third finger. Compare the size differences in these bones. Inset: The bony tendons of the tail run along most of the length of the tail. This is a diagnostic characteristic shared by all dromaeosaurs. It is similar to what is seen in the long tails of flying reptiles called pterosaurs. The similarity in the tails of both dromaeosaurs and pterosaurs is an example of convergence in which similar behavior led to unrelated animals having the same kind of physical characteristics.

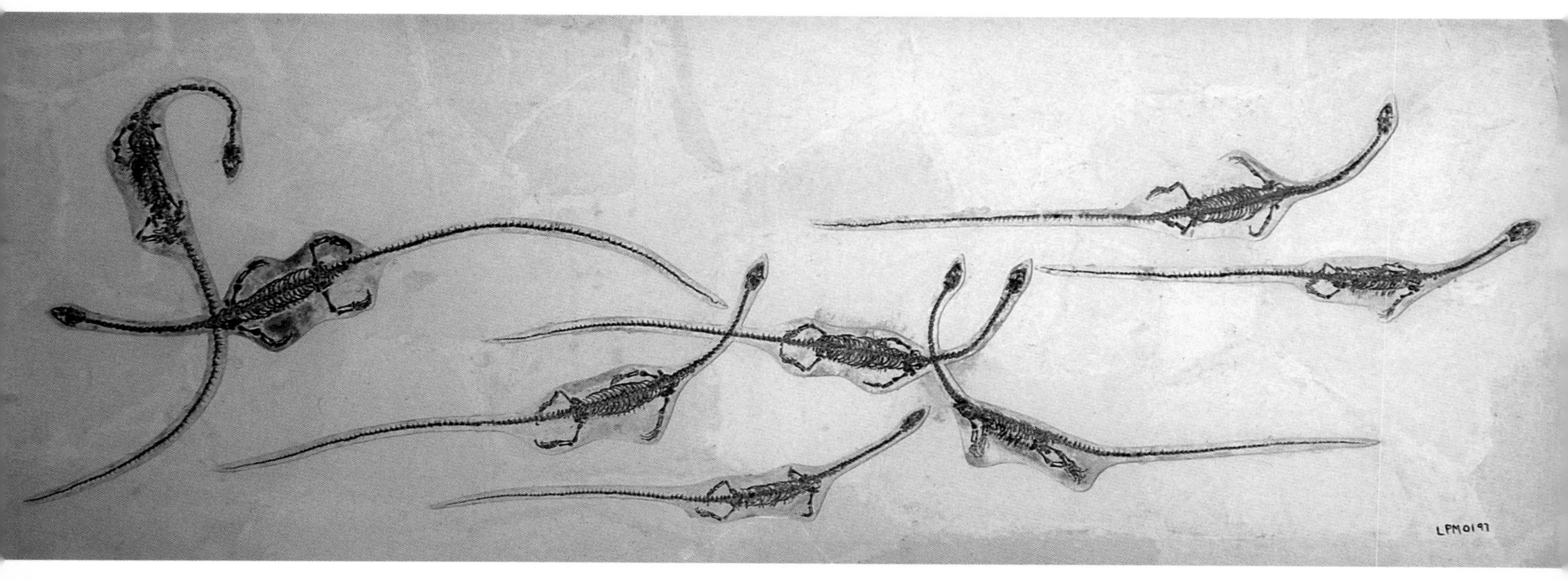

The *Sinohydrosaurus* seen above and alongside the dromaeosaur are aquatic reptiles. Just as their skeletons provide indications of their ability to swim, the skeletal anatomy of this dromaeosaur shows that it could fly.

The teeth and long bony tail of dromaeosaurs may look peculiar for a bird, but these are relics held over from their reptilian ancestry. As fossils of *Archaeopteryx* demonstrated so long ago, the flight feathers were unequivocal evidence that it was really a bird. The same can now be said of dromaeosaurs. And just as can be seen with *Archaeopteryx*, even without feathers preserved, there are unique characteristics in the skeleton of dromaeosaurs which reveal that they are truly avian.

Since *Archaeopteryx* represents an extremely archaic bird with a very poorly developed flight anatomy, comparison of the same bones used for flight in dromaeosaurs can be used in identifying that they had some ability to fly, and were also birds. The manner in which the hand folds back is unique in birds. The similarities seen in fossils of dromaeosaurs make sense when it is associated with having flight feathers on the hands. Subtle differences seen in the fingers also indicated that the hands were equipped with feathers. The middle finger in birds is where the flight feathers are attached. Birds after *Archaeopteryx* have a greater development of the middle finger which becomes thicker, or more robust as a greater support for the flight feathers. Look closely at the photo on the top left of the opposite page and you can see the second finger is slightly larger than the others. This is a strong indication that flight feathers were present. The flight anatomy of the chest and shoulders in dromaeosaurs are also strikingly similar to that in *Archaeopteryx*. But it is more developed in having a large sternum, or breastbone, made of bone. In *Archaeopteryx* the sternum is so undeveloped that it is missing probably because it was only cartilaginous.

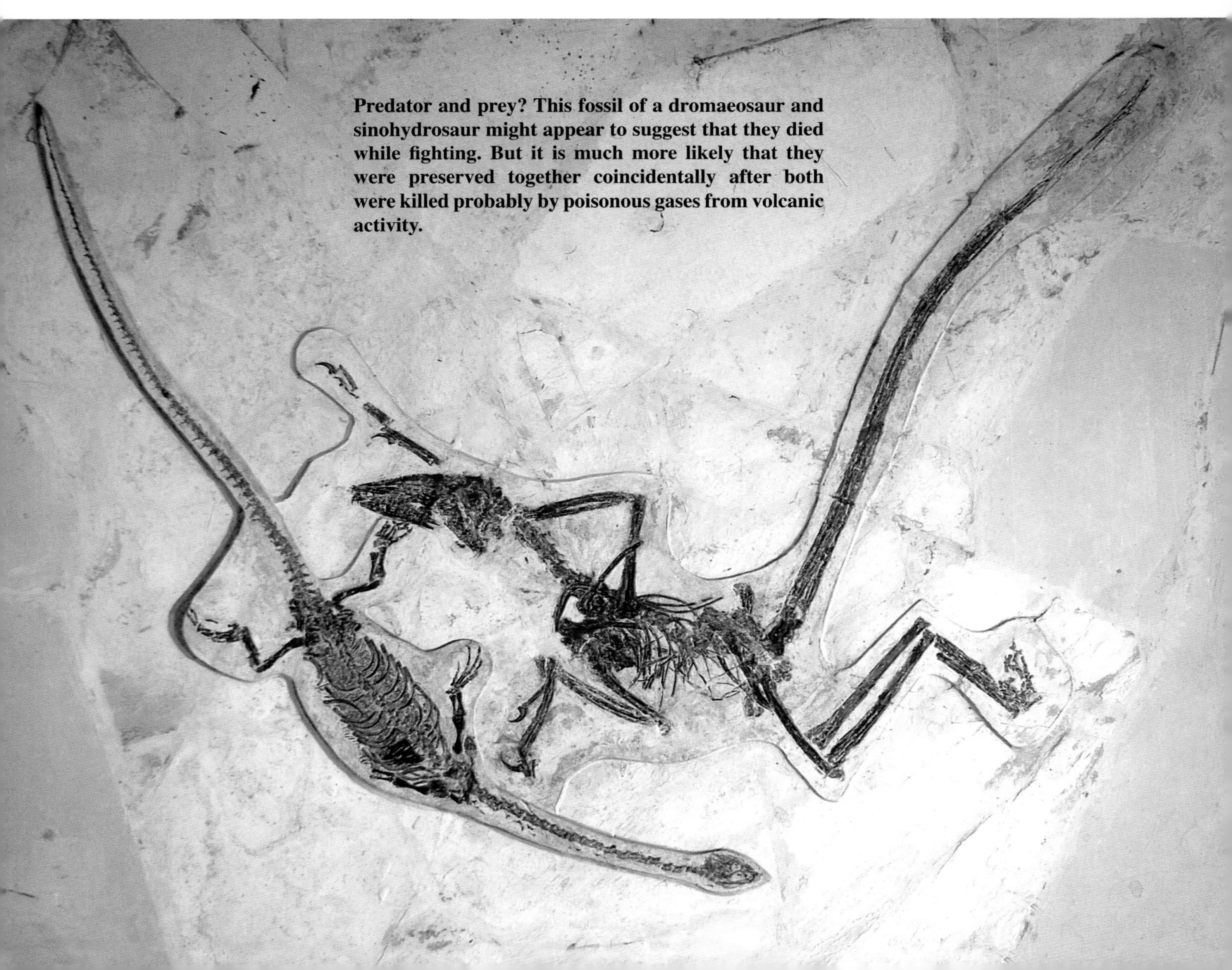

Predator and prey? This fossil of a dromaeosaur and sinohydrosaur might appear to suggest that they died while fighting. But it is much more likely that they were preserved together coincidentally after both were killed probably by poisonous gases from volcanic activity.

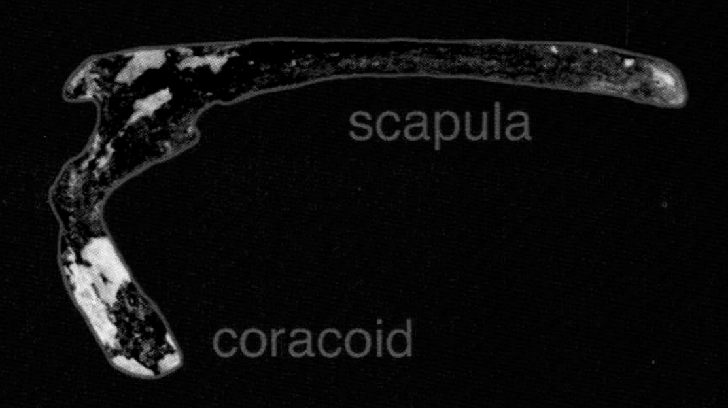

Flying Dromaeosaur

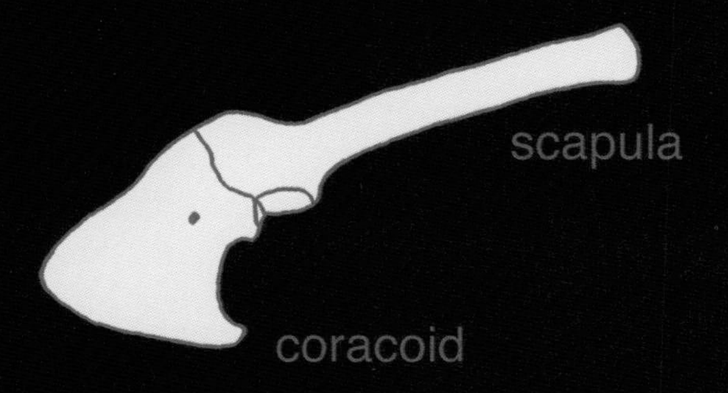

Flightless Dromaeosaur

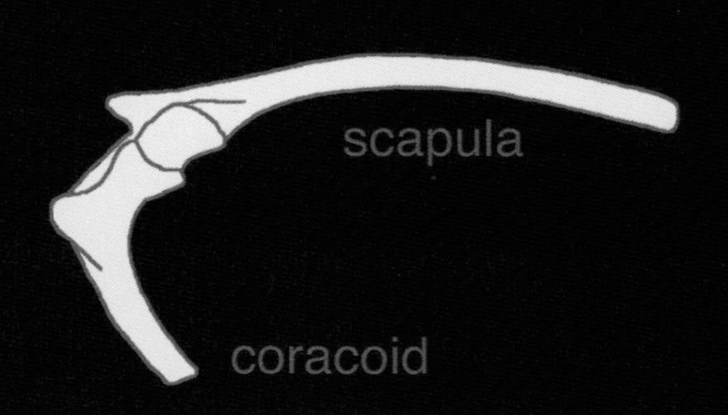

Archaeopteryx

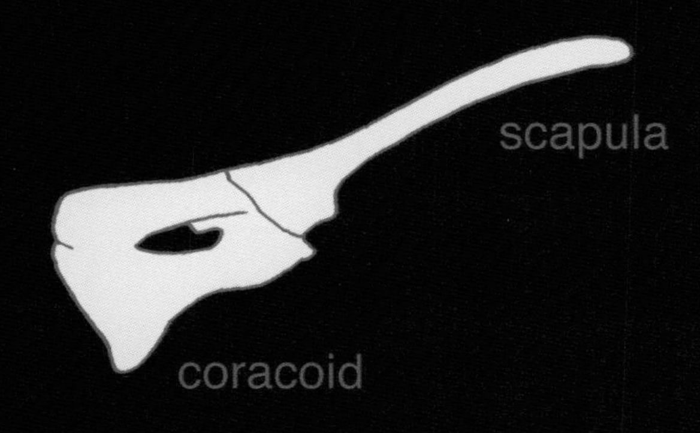

Ostrich

INDICATIONS OF FLIGHT AND FLIGHTLESSNESS

Even without feathers preserved, there are certain bones which can help determine if an animal was, or was not capable of flying. The scapula and coracoid are bones from the shoulder and chest which can reveal if the animal could fly. These bones in *Archaeopteryx* are oriented like those of modern flying birds which have the coracoid set at a right angle to the scapula. In comparison with *Archaeopteryx*, these bones from the dromaeosaur in the upper left strongly suggest that both were birds which could fly. Flightless birds like the Ostrich have coracoids which have reverted into being much more in line with the scapula. The scapula and coracoid in the figure on the top right is from the dromaeosaur, *Deinonychus*. The similarity to that of the Ostrich suggests that *Deinonychus* was also a secondarily flightless bird which had lost the ability to fly.

Dinosaur or bird?

Probably the most familiar portrayals of dromaeosaurs are those in motion pictures like *Jurassic Park*. Known as "raptors" these were based on *Velociraptor*, a dromaeosaur from Asia closely related to *Deinonychus* from North America. The evidence from the dromaeosaur fossils of Liaoning is a strong indication that the scaly depictions of raptors are not correct. Instead, they should be covered in feathers.

How different everyone's understanding of dinosaurs would be if only the dromaeosaur fossils of Liaoning had been discovered before *Velociraptor* or *Deinonychus*. If these smaller dromaeosaurs were known first, their identification as birds would have been easier to accept. In turn, this would have prevented larger forms like *Deinonychus* from being misinterpreted as bird-like dinosaurs. Having the precedent of flying dromaeosaurs already established would have helped identify larger dromaeosaurs as the birds they are. Their large size and avian attributes would make more sense as indications of being a flightless bird rather than being a dinosaur evolving towards becoming a bird. Most, if not all, of the debates on the origin of birds would probably never have happened, or at least would have transpired very differently. Had the feathered dromaeosaurs been discovered first, there would have been much more reason to believe bird-like theropods were actually birds instead of dinosaurs. It would be more evident that some "theropods" were actually birds which had lost their ability to fly and became secondarily flightless.

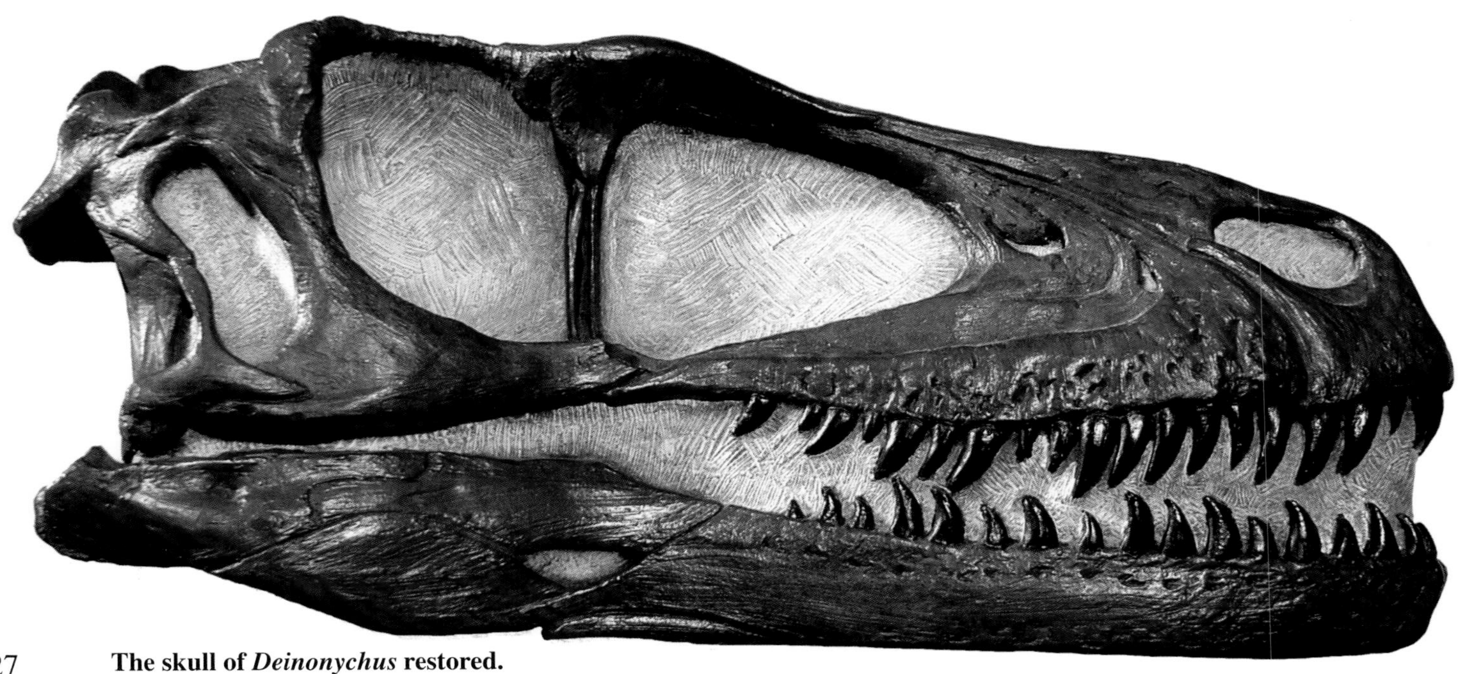

The skull of *Deinonychus* restored.

SCALY...

Instead of using speculative imagination to justify the possibility of feathers, the original depictions of *Deinonychus* were conservative and had a scaly hide as seen in mummified dinosaurs.

After more than three decades of having accepted the imagery of bird-like theropods as dinosaurs covered with scales, some may be hesitant to accept such a different interpretation of these animals as being so avian. However difficult as this may be, it actually simplifies understanding how birds evolved by taking away many of the controversial problems that have previously been so difficult to answer. Problems like explaining how or why ground dwelling dinosaurs would evolve bird-like forelimbs while not being able to use them as wings. Or trying to explain how very large bird-like dinosaurs got smaller and smaller until they could finally take advantage of their flight anatomy and learn to fly. These kinds of questions have been misleading because their premise is incorrect to begin with. In other words, it didn't happen that way. What the fossils of feathered dromaeosaurs suggest is that it was basically the other way around.

At least some animals that have been thought to be "bird-like dinosaurs" were actually descendents of early kinds of birds which had lost their ability to fly. In losing the ability to fly, their wings became smaller and there was a tendency for the animals to grow larger. This is very plausible because it is the same pattern which is known to have happened among many kinds of birds living today.

This different approach to looking at bird-like dinosaurs as being secondarily flightless birds is not only feasible, it is an important scenario which has been ignored for much too long. Indeed, it is very reasonable because the loss of flight is such a common occurrence among birds of recent times that it must have happened throughout the history of

birds. It may have even been more prevalent among the earliest birds because their flight anatomy was so weekly developed that it would have been easier to lose.

While the concept of some theropods being secondarily flightless birds has not been seriously considered or previously explored in depth by most researchers, this notable omission is starting to be rectified. A recent study has proposed that contrary to previous beliefs, larger bird-like theropods may have been descendents from smaller ancestors. This is a significant change in thinking because it acknowledges that the same pattern seen in secondarily flightless birds could apply to theropods.

The prospects that bird-like theropods like *Deinonychus* were really secondarily flightless birds is strengthened by the discovery of dromaeosaurs which had some ability to fly. This contradicts the concept that birds evolved from cursorial dinosaurs. It demonstrates that all of the avian characteristics seen in these potential avian ancestors can be accounted for in less complicated ways. In order to link dinosaurs to being the ancestors of birds, it has been cited that some theropods which are considered to be "non-avian" have over a hundred avian characteristics. However, the fossils of flying dromaeosaurs have exposed the real reason for them having so many avian characteristics is that they were already actual birds. So the evidence for dinosaurs as ancestors of birds is weakened considerably by the discovery of flying dromaeosaurs.

The question that all this poses then is if none of the bird-like dinosaurs actually represent the ancestors of birds, then what kind of ancestors did birds really have?

Mainslab and counterslab of *Scansoriopteryx* shown actual size.

THE EARLIEST BIRD

The inability to fly is not really sufficient criteria in searching for what kind of avian ancestors existed before *Archaeopteryx*. First, what is necessary is that they should be even less avian than *Archaeopteryx*. Then, there should be some indications towards the development of flight that are less developed than in *Archaeopteryx*. Most bird-like theropods have too many avian characteristics to represent what predated *Archaeopteryx*. What real ancestors of birds prior to *Archaeopteryx* should have is less of an ability to fly and fewer avian characters.

One such animal is *Scansoriopteryx heilmanni*. Its name means "climbing wing" and is in honor of Gerhard Heilmann who was the leading proponent of the concept that birds evolved from ancestors which could climb.

Scansoriopteryx is significantly different from all known dinosaurs. It had the ability to climb. Its uniquely designed hand was unlike any dinosaur in that it had an extremely long third finger. This is unlike the hands of theropods in which the third finger is shorter than the second.

It is very unlikely that such a long finger could have evolved from the condition seen in carnivorous dinosaurs. This is because, usually as theropods evolved, there was a greater tendency for the fingers to get shorter instead of longer. What this suggests is that *Scansoriopteryx* appears to reflect an earlier ancestry than that of dinosaurs when the third finger was still longer than the other two. This is a startling implication, but other parts of the skeleton also indicate that *Scansoriopteryx* evolved from along a different evolutionary path than dinosaurs.

The unique design of the hand would not have been very practical for grabbing onto prey because the long third finger was not very flexible. It would have also been very prone to injury because it stood out so far from the rest. However, by sticking out so far, it would have given the three claws of each hand a broader grasp making it very effective in catching onto twigs or branches. Imagine if *Scansoriopteryx* jumped from one branch to another. Having the kind of hand that it did would have been very helpful in making sure that it could grab onto things and be able to make a safe landing.

There is reason to believe that *Scansoriopteryx* could have used all the help it could get to make safe landings. It had a flying ability that was even less than that of *Archaeopteryx* which is believed to have been a very poor flier. At best, it probably did little more than jump and glide. None of its flight anatomy was as well developed as in *Archaeopteryx*. Its long arms were definitely much less bird-like. So much so, that if not for a single bone it would not be possible to recognise it as a wing at all. This bone is the semilunate carpal, a half-moon shaped bone in the wrist that is unique to birds.

In *Scansoriopteryx*, the semilunate carpal is the only bone which conclusively ties it to being a bird. Without this particular bone there would be no evidence that it had any ability to fly. What the semilunate does is allow the hand of a bird to move in a very restricted sideways range of motion which is crucial in being able to stabilizing itself while flying. It also makes it possible for birds to fold the hand and primary feathers backwards along the sides of the body. There is no reason why *Scansoriopteryx* would have been equipped with such a well developed semilunate carpal unless it had feathers on its hands and at least some ability to fly, or at least glide. So regardless of it being less avian than *Archaeopteryx* in any other details, there is just enough evidence in the skeleton of *Scansoriopteryx* to qualify it as a bird.

Above: *Scansoriopteryx* nestlings react to something approaching. Below: An enlarged detail of the hand from the mainslab of *Scansoriopteryx*.

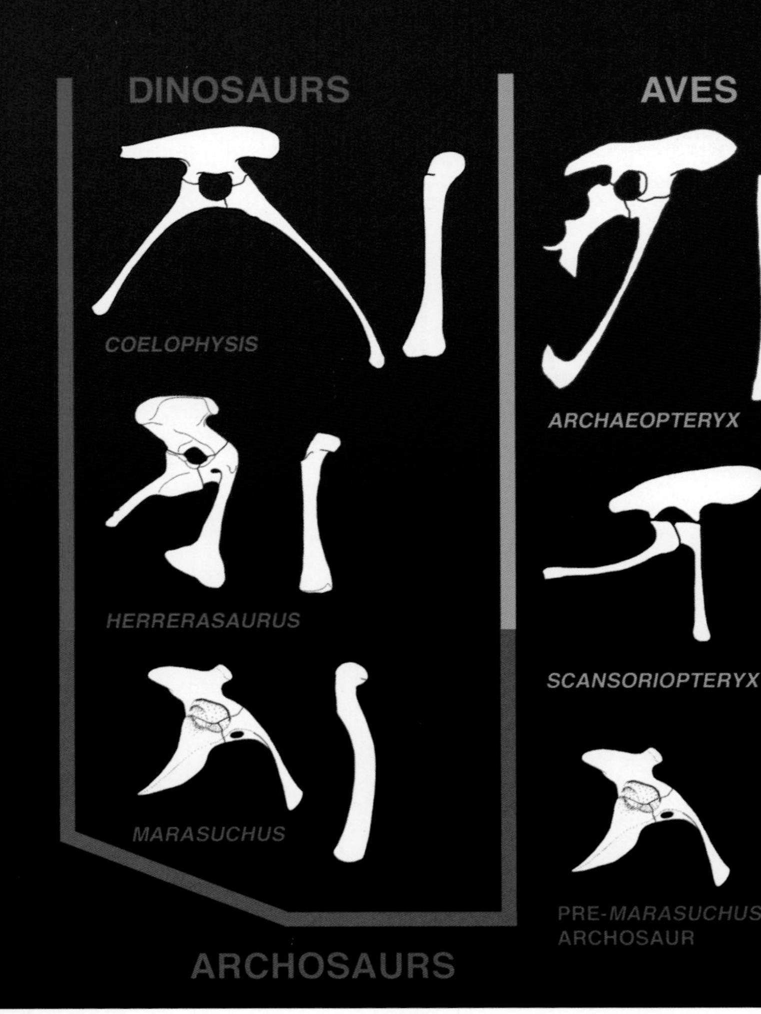

SEPARATE LINEAGES

Dinosaurs and birds evolved from archosaurs, but from different ancestors. In the bird *Scansoriopteryx*, the structure of the leg is unlike those of dinosaurs in not having an offset head of the femur. This is a more ancient condition that would have existed in some archosaurs before *Marasuchus*. What this means that is that birds did not evolve from dinosaurs. Instead, *Archaeopteryx* must have evolved the inturned head of the femur independently of dinosaurs. In birds, the hole in their hip sockets would have also evolved separately from dinosaurs. These similarities between birds and dinosaurs are not evidence of a shared heredity but are instead examples of convergence.

Compare the undeveloped head of the femur in *Scansoriopteryx* to that of the baby theropod.

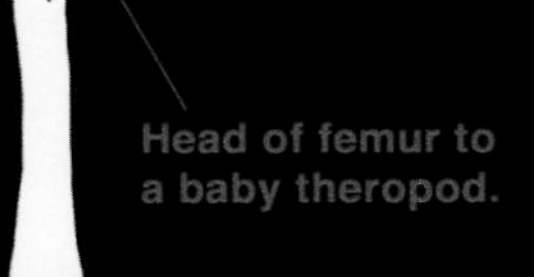

Head of femur to a baby theropod.

The head of the femur to a baby theropod is typical of dinosaurs. This indicates that the non-dinosaurian leg structure of *Scansoriopteryx* is not due to an immature stage of growth.

Scansoriopteryx qualifies as being a bird, but does it qualify as a dinosaur? There is reason to think that it does not.

Among the most basic criteria for being a dinosaur is its upright limb posture. It has generally been presumed that all dinosaurs are descendents of a bipedal ancestor. This ability to stand up and move about on only two legs is reflected by special modifications in the pelvis and legs. In dinosaurs, the inner back side of the hip socket has an opening where it is normally solid in reptiles. The head of the femur fits into the hollowed out hip socket. But what is more significant in creating the upright stance in dinosaurs is that the head of the femur is offset at a right angle which basically repositions the hind leg into a vertical posture. All birds from *Archaeopteryx* on have this kind of special anatomy which makes it possible to be bipedal. So this alone might appear enough to justify regarding all birds as dinosaurs.

There is a problem though. A serious problem. The femur of *Scansoriopteryx* does not have the articulating head offset any more than what is seen in reptiles with legs that sprawl out to the sides. The hip socket has a small opening, but not exactly what should be expected of a real dinosaur. How can this be if birds evolved from dinosaurs? The answer is they did not.

Since *Scansoriopteryx* did not have the full development of the upright limb posture, it could not have evolved from dinosaurs which already did. What this means is that the upright limb posture seen in birds and dinosaurs must have evolved independently of each other. Therefore, this evidence means that birds and dinosaurs represent separate lineages.

It is remarkable that *Scansoriopteryx* should have characteristics which not only disqualify it as a theropod, but also as even being a dinosaur. Still, other characteristics in the skeleton corroborate that *Scansoriopteryx* is derived from such an ancient ancestry that it must have predated the earliest true dinosaurs.

Unlike *Archaeopteryx* which has a more avian pelvis in which the pubis points backwards, the same bones of *Scansoriopteryx* point forward in the typical reptilian style which indicates that it was from an earlier lineage. In *Scansoriopteryx* the pubic bones are so short that they compare best with what is often regarded as the ancestors of dinosaurs, archosaurs like *Marasuchus*. But despite this similarity, even *Marasuchus* had a greater offset articulation in the head of the femur than in *Scansoriopteryx*. So even at the ancient stage of dinosaurian ancestry represented by *Marasuchus*, the avian lineage was still separate and even older.

The idea that birds are not really dinosaurs may appear rather heretical in light of how widely popular it has become to regard birds as "living dinosaurs". Of course, it's fun to think of birds as dinosaurs. But actually, what *Scansoriopteryx* represents is the traditional version of the origin of birds that had been believed for most of the past century.

If only *Scansoriopteryx* been discovered back in the days of Heilmann, he would have had the evidence to substantiate the speculation that birds evolved from arboreal ancestors which were not dinosaurs. Through the years, many scientists have continued to believe that birds were descended from some unknown branch of archosaurs basically as Heilmann had predicted. But the mistaken identity of dromaeosaurs led to the belief that birds had evolved from ground dwelling dinosaurs. In turn, this led to the belief that there was no such thing as the arboreal ancestor of birds. It was thought to be a myth, an animal that simply did not exist. But now that it has been discovered, *Scansoriopteryx* changes the dynamics of the entire argument. Birds did have an arboreal ancestry. And birds are all the more special because their ancestry is even older than that of dinosaurs.

X-ray of a *Confuciusornis* fossil.

A DIVERSITY OF BIRDS

The fossils of birds have always been among the rarest of fossils. Certainly, most bird fossils are few and far between being known by only one specimen, or perhaps a few at best. The exception to this is *Confuciusornis*. Not only dozens, but hundreds of these birds have been discovered. The quality of preservation varies considerably, but there is much that can be learned from this spectacular array of so many of the same kinds of prehistoric birds.

The variation in sizes can be associated with different growth stages. Differences in the kind of feathers used for ornamentation can distinguish the males from the females. It appears that the males had two long tail feathers, whereas the females did not. Like so many birds of today, *Confuciusornis* was probably very social and, at least at times, congregated in large flocks. It is not too uncommon to find two or more individuals preserved together. All in all, the lifestyle of *Confuciusornis* might not have been dramatically different from that of modern day birds, and yet despite any similarities, they were significantly different from their living counterparts in many ways.

As seen in the x-ray on the previous page, the skeletal anatomy of *Confuciusornis* is clearly that of a bird. It has so many avian characteristics that any number of individual bones could be used to distinguish it as a bird. Unlike specimens of *Archaeopteryx* or dromaeosaurs which sometimes have not been properly recognized as birds, there is no problem

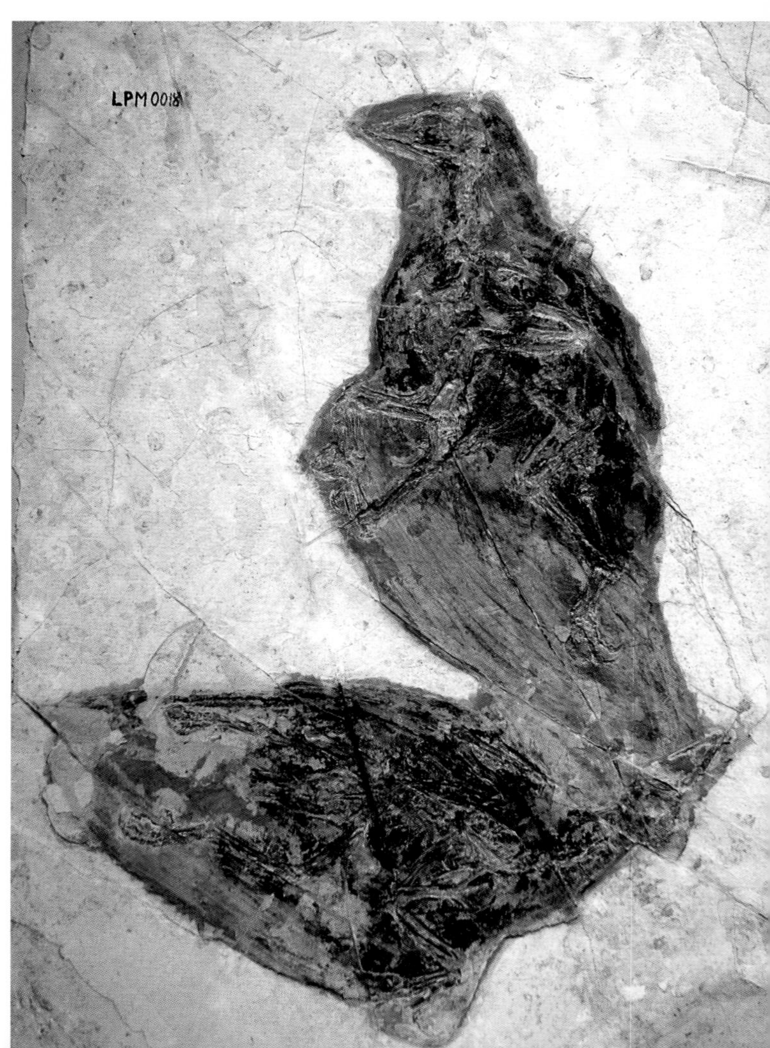

in being able to determine that *Confuciusornis* is a bird from just looking at the skeleton. The preservation of feathers is not necessary. Compared to *Archaeopteryx*, the flight anatomy is much more developed and is well on the way to resembling that of typical birds of today. The same bones had the same functions for flying, but they were still different in not being as perfected as they would eventually become in present day birds. Some archaic traits like having a long bony tail were lost and replaced with a stiffened pygostyle. Even the teeth had been entirely lost making the mouth look remarkably modern. But there is no doubt that *Confuciusornis* was still a very archaic type of bird representing a transitional stage of avian evolution. Unlike the prominent keeled sternum which normally supports large flight muscles, the breastbone of *Confuciusornis* was still relatively small and very poorly developed. At best, it had only a very small keel to support comparatively weak flight muscles. The furcula, or wishbone, was hardly like typical birds of today. It was relatively small and poorly developed, but it was significantly larger than in *Archaeopteryx*. The most obvious indications of its prehistoric ancestry was that it still retained three clawed fingers. While this is basically similar to *Archaeopteryx*, special modifications of the middle finger show that it had gotten much more robust to provide better support for the flight feathers.

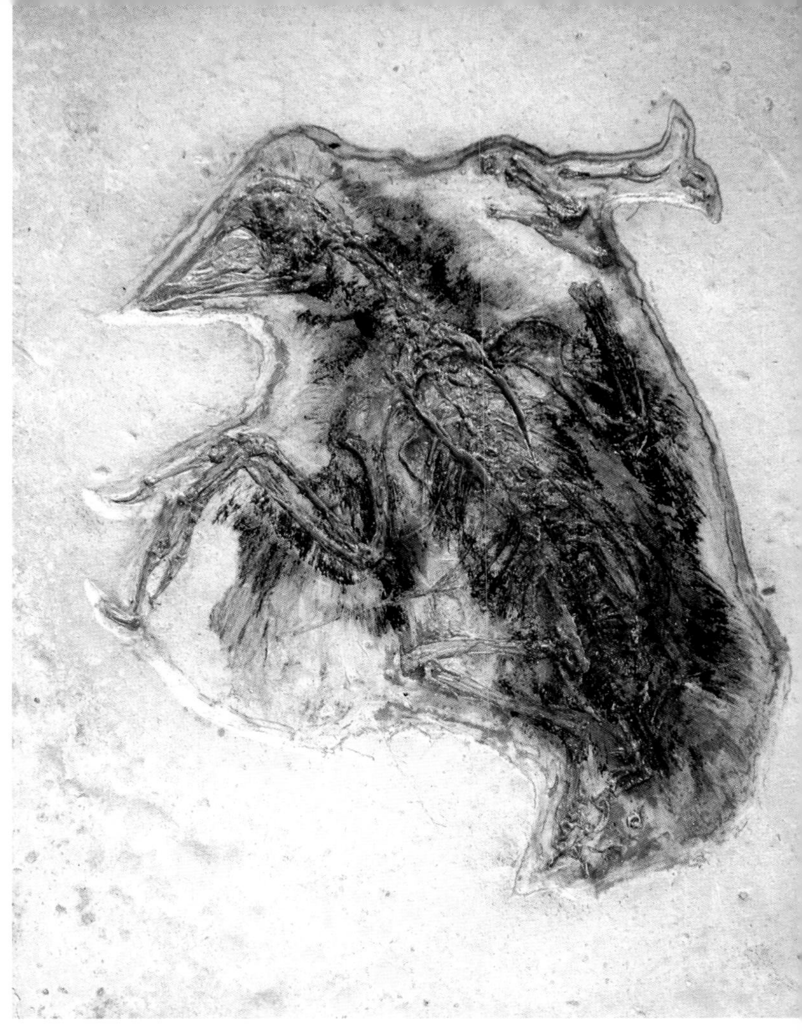

Note that the *Confuciusornis* seen above does not have its flight feathers preserved. This type of incomplete preservation often happens among the various kinds of birds from Liaoning, including those of dromaeosaurs. However, the flight feathers must have been there in life.

Opposite page: Faint and imperfect feather impressions leave a ghostly outline in this specimen of the tiny *Liaoxiornis*. Magnified approximately 3 times in size.

Left: The skeleton of *Eoeantiornis* without feathers displays many avian characteristics which are readily identifiable. Reduced to about half natural size.

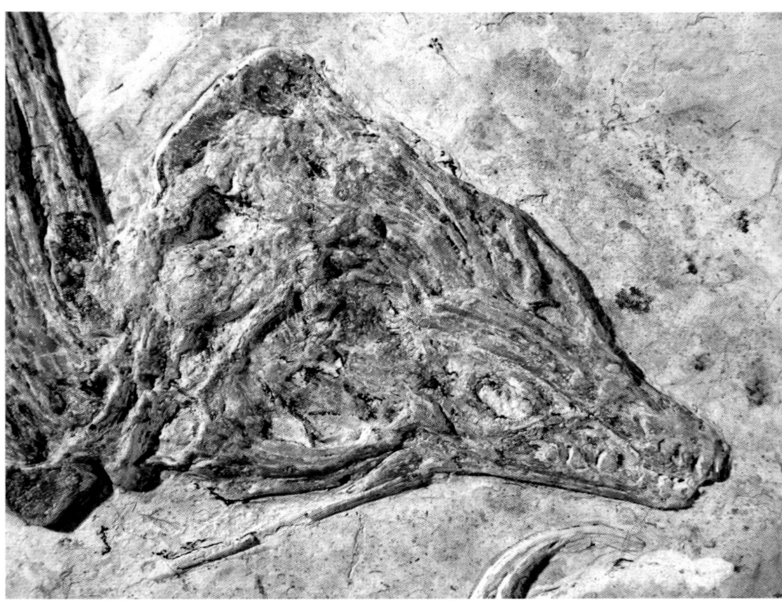

Above: The skull of *Eoeantiornis* reveals that it still has teeth in the front of its mouth.

Below: The furcula, or "wishbone", of *Eoeantiornis* has the distinctive "Y" shape seen present-day birds.

The toothless beak of *Confuciusornis* stands out from most other birds from the Early Cretaceous of Liaoning. It was much more common for birds known from this time, 125 million years ago, to have still retained at least some teeth. Even though the first teeth birds usually lost were from the back of the jaws, the beak almost certainly evolved by growing backwards from the tip of the snout. This suggests that the development of the beak began before all the teeth were lost. A keratinous beak would have extended from the front overlying any teeth and probably continued to cover the toothless part of the jaws.

It may seem odd, but many of the toothed birds, including *Liaoxiornis* and *Eoeantiornis*, had a greater developed flight anatomy and resembled modern birds more than did the fully beaked *Confuciusornis*. This means that different kinds of birds had their own developmental rate, or speed, in which avian characteristics were acquired. When compared to *Confuciusornis*, the hands and furcula of *Liaoxiornis* and *Eoeantiornis* are more like modern birds. Like birds of today, the third finger had degenerated to only a vestige of itself and was nearly lost. And the furcula had taken on the more familiar shape of a typical modern wishbone.

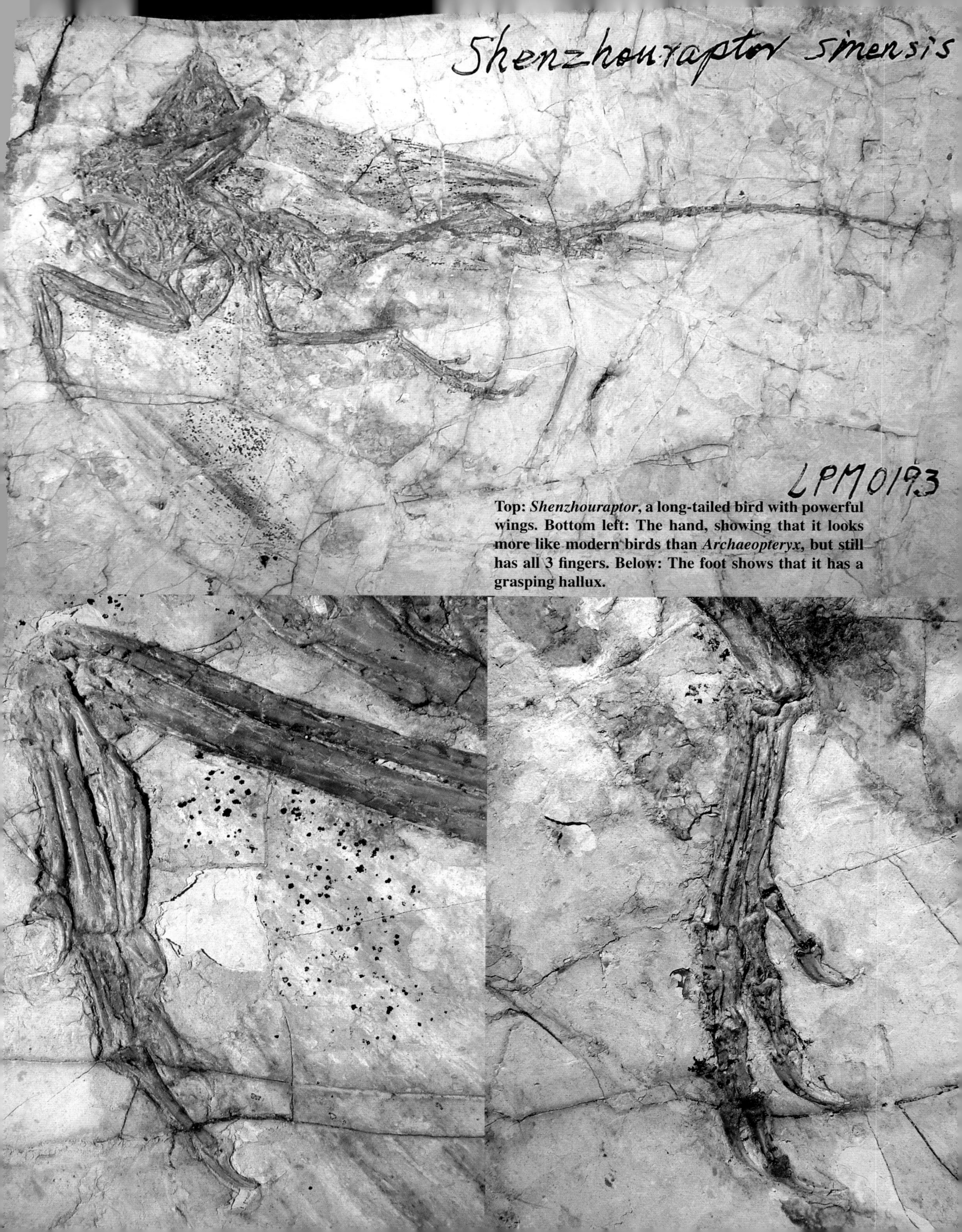

Top: ***Shenzhouraptor***, a long-tailed bird with powerful wings. Bottom left: The hand, showing that it looks more like modern birds than ***Archaeopteryx***, but still has all 3 fingers. Below: The foot shows that it has a grasping hallux.

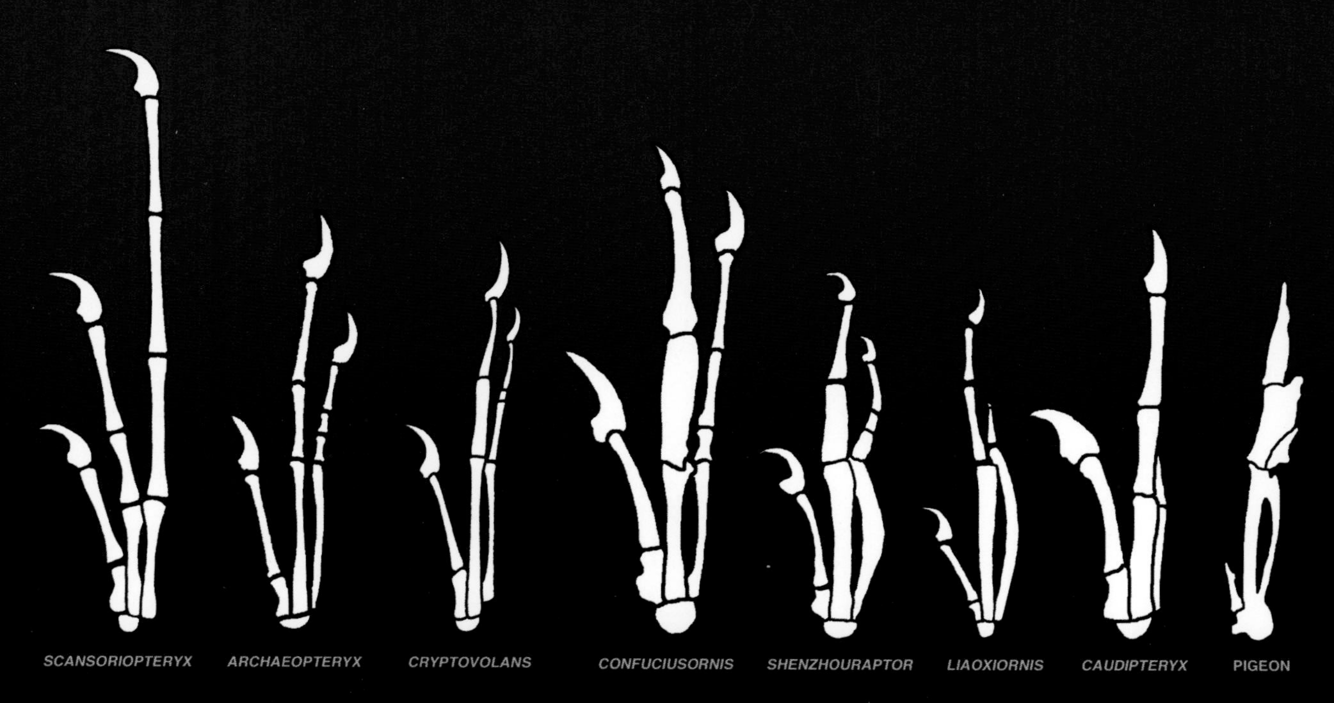

Seen above is an evolutionary chain in which the hands of birds went from having specializations for climbing, to the reduction of finger size, followed by the loss of the third finger and claws.

It is a common mistake to think that all birds after *Archaeopteryx* had shorter tails. There was obviously a strong tendency to lose the long bony tail and develop the short pygostyle, but there were exceptions. The dromaeosaurs were more powerful in their flight anatomy than *Archaeopteryx*. And yet, the tails of dromaeosaurs remained long in spite of the wings becoming more powerful. Another type of long-tailed bird is *Shenzhouraptor*. It also retained a long bony tail even though its wings had become considerably larger and more powerful than other birds with weaker wings and short tails.

It is very significant that long-tailed birds continued to exist even though they had developed stronger flight anatomy and more modern looking wings. One type of long-tailed bird, *Rahonavis*, continued to exist to nearly the end of the age of dinosaurs. This means that long-tailed birds coexisted throughout most of the Cretaceous with short-tailed birds. This is an exceptionally long period of time to coexist, if there really was a competitive edge in having short tails. This suggests that it may be an oversimplification to attribute the eventual extinction of long-tailed birds to the superiority short-tailed birds. Other factors, still unknown, must have contributed to why long-tailed birds eventually went extinct.

For most of the past century, there have been only a few examples to show various stages of how the wings of birds evolved from *Archaeopteryx* to those of modern-day birds. The fossils from China have contributed greatly to being able to understand how this process occurred. The hands of birds exhibit successive changes as they developed their ability to fly. The earliest stage, represented by *Scansoriopteryx*, had the longest outer, or third finger. By the time *Archaeopteryx* appeared, the third finger had become shorter than the second. The hands of *Cryptovolans* and *Microraptor* resemble those of *Archaeopteryx* but the first phalange of the middle finger had become slightly thicker which allowed a stronger attachment for the feathers. The hand of *Confuciusornis* still retained three fingers, but the first phalange of the middle finger has become much thicker for better support of the feathers. The claw of the middle finger is degenerative representing the process of eventual loss common to most modern day birds. The third finger in the hand of *Shenzhouraptor* is markedly reduced in size but only hints at the evolutionary process in which the third finger continued to degenerate in birds. In *Liaoxiornis*, the third finger is only a vestige retaining just a single phalange which is like the condition seen in birds living today, such as the common pigeon.

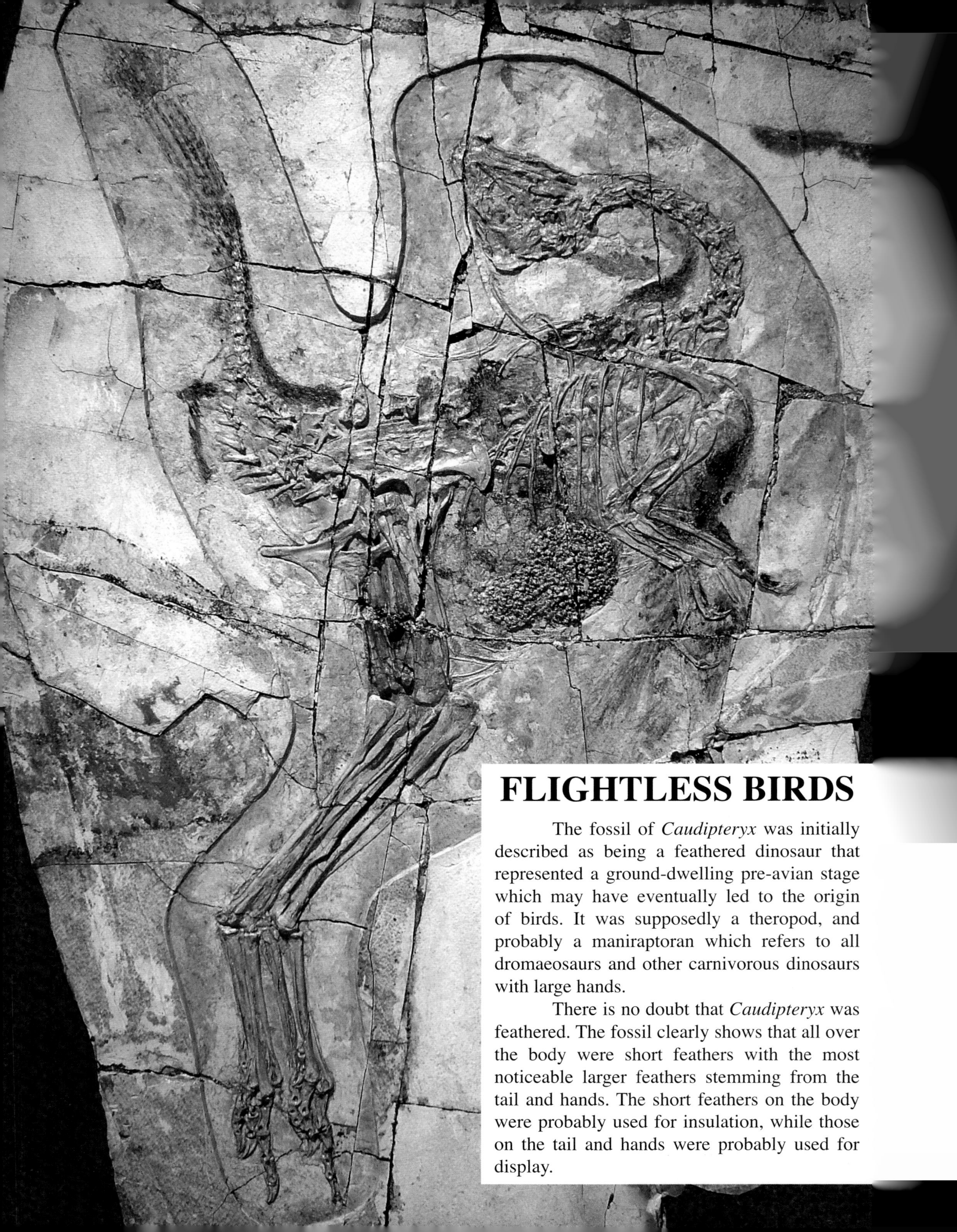

FLIGHTLESS BIRDS

The fossil of *Caudipteryx* was initially described as being a feathered dinosaur that represented a ground-dwelling pre-avian stage which may have eventually led to the origin of birds. It was supposedly a theropod, and probably a maniraptoran which refers to all dromaeosaurs and other carnivorous dinosaurs with large hands.

There is no doubt that *Caudipteryx* was feathered. The fossil clearly shows that all over the body were short feathers with the most noticeable larger feathers stemming from the tail and hands. The short feathers on the body were probably used for insulation, while those on the tail and hands were probably used for display.

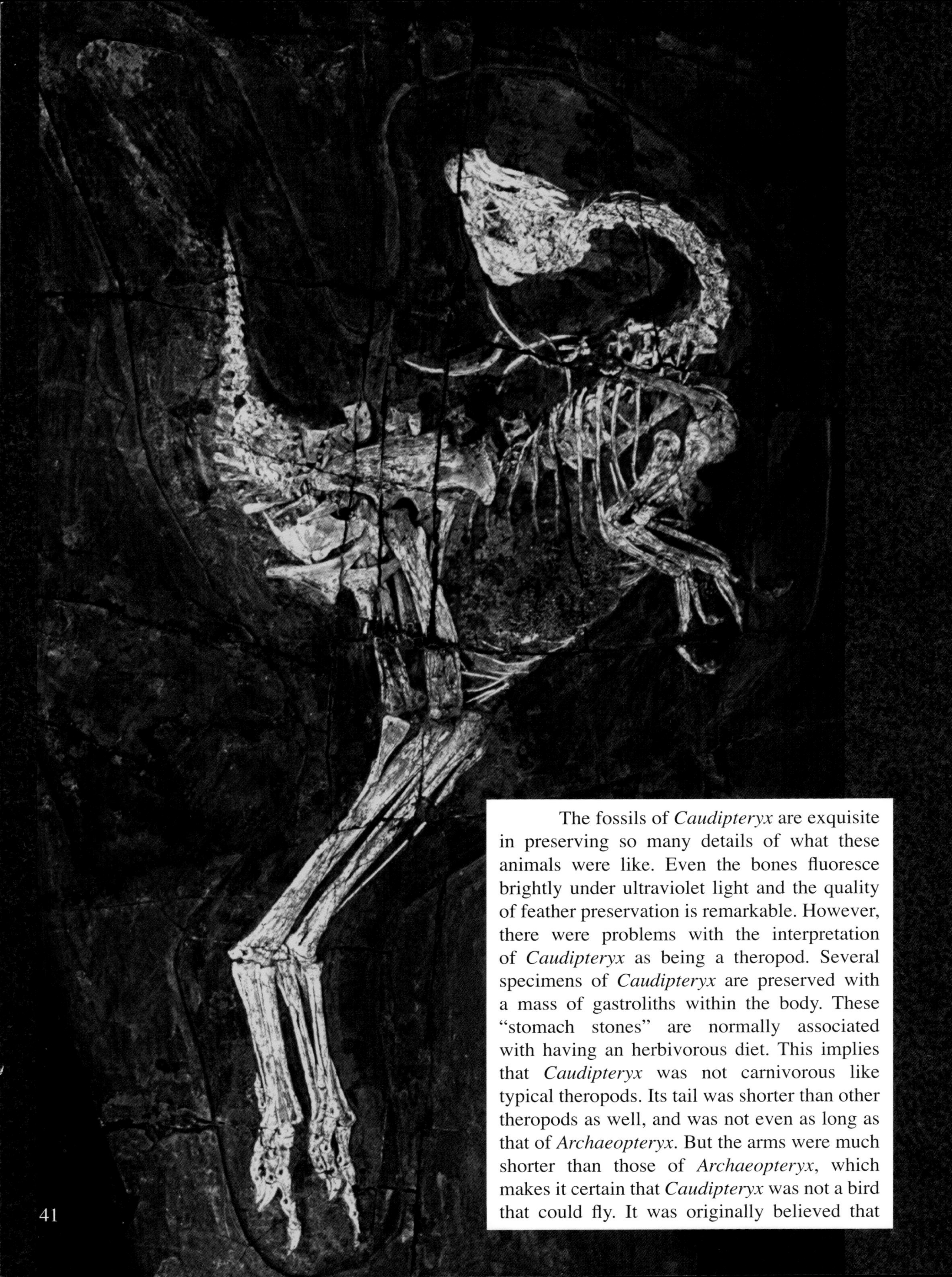

The fossils of *Caudipteryx* are exquisite in preserving so many details of what these animals were like. Even the bones fluoresce brightly under ultraviolet light and the quality of feather preservation is remarkable. However, there were problems with the interpretation of *Caudipteryx* as being a theropod. Several specimens of *Caudipteryx* are preserved with a mass of gastroliths within the body. These "stomach stones" are normally associated with having an herbivorous diet. This implies that *Caudipteryx* was not carnivorous like typical theropods. Its tail was shorter than other theropods as well, and was not even as long as that of *Archaeopteryx*. But the arms were much shorter than those of *Archaeopteryx*, which makes it certain that *Caudipteryx* was not a bird that could fly. It was originally believed that

Instead of having to regard animals like *Caudipteryx* as a feathered dinosaur, there is just as much reason, if not more so, to consider it as a flightless bird. Because some birds must have lost their ability to fly time and again since *Archaeopteryx*, or even earlier, it is important to ask why their fossils have not been discovered? The answer is that they may have, but these flightless birds have been misidentified as dinosaurs.

Caudipteryx had three clawed fingers like most theropods. But this was incorrect. The third finger was degenerate to the point that the hand more closely resembles that of a modern-day bird. So it should come as no surprise that there has been considerable difficulty in properly identifying what *Caudipteryx* really represents.

There needs to be a logical explanation as to why *Caudipteryx* had the long feathers on its hands. Is using them for display really a sufficient answer? Perhaps. But should this mean that these feathers originated for the sole purpose of display? Not necessarily. If *Caudipteryx* was a bird which had lost its ability to fly, this would explain why it had the kind of feathers that it did. It may have had a flying ancestor which had asymmetrical flight feathers. If so, then as seen by the wings of ostriches and other ratites, the flight feathers of *Caudipteryx* could have reverted back to being symmetrical and incapable of flight.

A significant clue as to how flight feathers first evolved can be found in how they grow in the hatchlings of birds. It may even help answer what caused the semilunate wrist bone to develop in birds. As newly hatched chicks grow, their bodies are covered with down feathers to help keep them warm. The first signs of flight feathers appear on their hands before they do on the rest of the arms. If this pattern of development reflects what happened in the evolution of birds, then what the earliest birds evolved first in the structure of the wings would have been the flight feathers stemming from the hands. This would contradict the speculative depictions of bird-like theropods which show flight feathers from the shoulder joint down to the wrist, but not on the hands. This is important because such depictions do not explain why the semilunate would have evolved. If, however, the flight feathers first appeared on the hands of the earliest birds, then as they grew larger they would have continually increased their aerodynamic forces. As the flight feathers evolved into longer structures, the movement in the wrist would have been naturally restricted into more of a sideways action than the usual up and down range of motion. By bending the wrist back and forth while the hand and flight feathers were extended out horizontally, there would have been an ever better ability to control the aerodynamic stability required to safely move through the air. So instead of existing for some unknown reasons unrelated to flight, feathers on the hands could have directly caused the semilunate wrist bone to evolve.

The discovery of wing feathers on dromaeosaurs suggest that depictions of bird-like theropods have been incorrect in not having feathers on the hands.

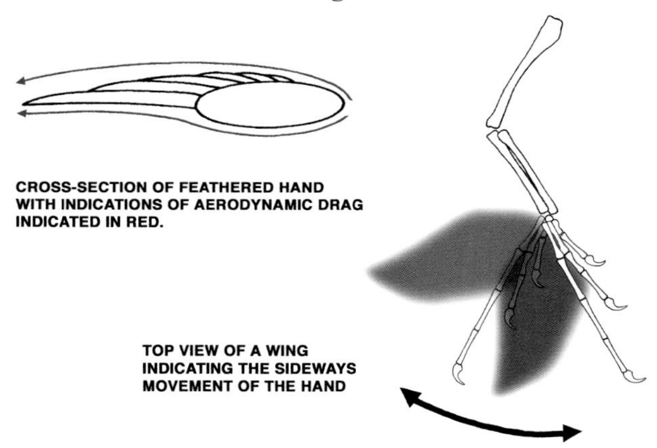

The earliest progenitors of bird wings may have had feathers elsewhere along the length of the arm, but the presence of even small feathers on the hands would have created aerodynamic forces that could have naturally contributed to the development of the semilunate carpal. The red arrows in the figure on the left represent aerodynamic drag which could be controlled by the sideways motion of the hand.

As seen above in the hatchling of a chicken, the flight feathers of the hands are the first to develop on the wings. This may be a clue as to how wings evolved.

CONCLUSIONS

The question as to how birds are related to dinosaurs is complex. Finding the answer has been complicated by mistaken interpretations and missing important pieces of the puzzle. The fossils from Liaoning have revealed significant information which contradicts the popular view that birds of today can be considered the last living dinosaurs because they had supposedly evolved from theropods. Fossils of animals that were definitely not supposed to be birds have turned out to be just that--birds. The basis for thinking that these birds were dinosaurs depended on them not having flight feathers or any ability to fly. That the arms of dromaeosaurs have actually turned out to be wings, complete with flight feathers indistinguishable from modern-day birds which fly, demonstrates that thinking these animals were dinosaurs is not correct. Instead of proving that birds and dinosaurs were related, what this revealed is that birds had been misidentified as dinosaurs. Then, this mistake was compounded by using these birds, misinterpreted as dinosaurs, to represent what the ancestors of birds were supposedly like before flight had been achieved.

It is important to make the distinction that dromaeosaurs were not dinosaurs on their way to becoming birds, but instead were birds which had been misidentified as dinosaurs. This idea dramatically alters the scientific argument of how birds supposedly evolved the ability to fly from ground-dwelling animals. But as devastating as this is to the concept that birds evolved from dinosaurs, the discovery of *Scansoriopteryx*, an arboreal bird which preceded *Archaeopteryx*, is even more overwhelming. The traditional belief that birds were derived from climbing ancestors so ancient that they predated dinosaurs is once again the more viable interpretation.

Convergence, the evolution of similar characteristics among unrelated animals occurred throughout the history of dinosaurs and birds. While the issue of convergence has been complicated by the misidentification of

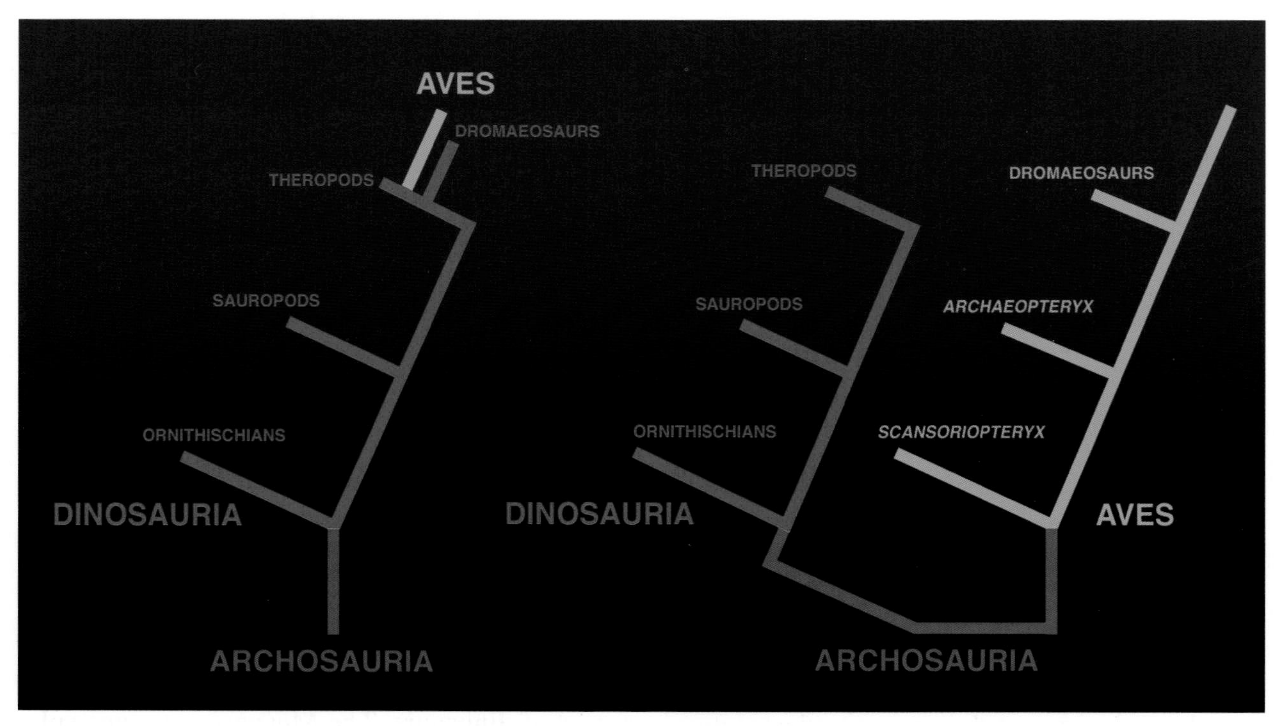

The two charts shown above depict conflicting interpretations of how birds evolved. In the chart on the left, birds are regarded as dinosaurs having evolved from theropods after dromaeosaurs. The chart on the right illustrates how birds and dinosaurs evolved from separate lineages that stem from different ancestors from within the Archosauria. The ancestry of the avian lineage stems further back into the Archosauria than that of the Dinosauria. Note that in both charts, Aves leads to all birds.

Standing over 14 feet tall, *Therizinosaurus* was probably a giant flightless bird and not a feathered dinosaur.

birds as theropods, it needs to be pointed out that not all theropods were birds. Only those which had the ability to fly in their ancestry should be considered truly avian. This would include giants such as *Therizinosaurus* which was derived from ancient birds that had lost their ability to fly.

The fossils of Liaoning have not only revealed mistakes in previous interpretations regarding the origin of birds, they have also contributed to finding new answers. No doubt many of these fossils will be highly debated, scrutinized meticulously, and remain controversial. This is all part of the scientific process in the search for a better understanding. What is important is that the evidence is continuing to make the answers more clear and accurate. That dromaeosaurs and other bird-like "dinosaurs" have turned out to be birds does not adversely impact the concept of evolution. To the contrary, it strengthens the evidence for evolution by demonstrating how it worked. The idea that birds may not technically deserve to be regarded as dinosaurs may be disappointing to some. However, that birds are derived from an ancient lineage even older than dinosaurs is just as awesome if not more so. There is now strong evidence that birds evolved from arboreal ancestors, instead of ground dwelling theropods. This means the scientific argument has been brought back to what was traditionally accepted for most of the past century.

The different developmental stages in the flight anatomy represented among various avian fossils clearly demonstrate how birds evolved their ability to fly. The sequential changes in the skeletal structure represent the process of evolution itself. The discovery of more fossils from around the world, and in particular from Liaoning, will certainly continue to contribute even more to the current state of knowledge. But already, a much more complete picture has taken shape regarding the origin of birds and the world of dinosaurs. The Mesozoic was not only the great Age of Reptiles, it was a world filled with birds that continued to evolve more efficient anatomy for flying, as well as, numerous kinds of birds which had lost their ability to fly.

GLOSSARY

Arboreal - Adapted with climbing ability for living and moving about in trees.
Archosauria - A classification of reptiles with unique openings in the sides of their skull which includes many extinct forms, as well as, the ancestors that led to dinosaurs, crocodilians and birds.
Aves - The classification that includes all birds.
Avian - Of or pertaining to birds.
Bipedal - Moving on two legs.
Bird - All vertebrates that have feathers used for flying, and subsequent forms which have lost the ability to fly.
Calamus - The root of a feather.
Convergence - Refers to anatomical characteristics that look similar even though they evolved independently.
Coracoid - A bone of the chest in birds and reptiles that connects with the scapula and sternum.
Cretaceous - The period of geologic time during the Mesozoic era between 144 and 65 million years ago.
Cursorial - Running ability.
Dinosauria - Archosaurian reptiles that have a fully perforated acetabulum (hip-socket) and have the head of the femur offset at a distinct right angle. Includes the Ornithischians which have retroverted pubes and Saurichians which have pubic bones oriented forwards. Does not include Aves, or birds.
Dromaeosaurs - Birds that have long tails stiffened by bony tendons. Includes flying and secondarily flightless forms. Popularly known as "raptors". They were previously regarded as bird-like theropods and non-avian dinosaurs.
Evolution - The natural process of change from one form to another related form.
Femur - The upper bone of the leg.
Flight feathers - Feathers that have asymmetrical vanes.
Flight anatomy - Refers to the characteristics specifically associated with having the ability to fly.
Fossil - Any structural remains, impressions, or traces of animals or plants preserved from the prehistoric past.
Furcula - The characteristic "wishbone" of birds. Also exist in some theropods.
Hallux - Digit 1, or the first toe, in birds which can be oriented backwards as an adaptation for perching or grabbing prey.
Maniraptoran - A classification initially used for "bird-like, non avian theropods" with large hands, but now may be included within Aves.
Mesozoic era - The Age of Dinosaurs and the geologic period of time 245 to 65 million years ago, consisting of the Triassic, Jurassic and Cretaceous.
Pennaceous feathers - Feathers that have flat, sheet-like vanes with interconnecting barbs.
Plumulaceous feather - Feathers that have loose barbs and fluffy vanes.
Pygostyle - The shortened tail of a bird formed by fused vertebrae.
Quadrupedal - Moving on four legs.
Rachis - The central stem of a feather.
Scapula - Shoulder blade.
Secondarily flightless - The loss of the ability to fly.
Semilunate carpal - The unique half-moon shaped bone connected to the base of the hand in birds that restricts the movement of the wrist in a sideways range of motion and used for flight.
Shoulder girdle - Combination of the scapula and coracoid bones.
Sternum - The breastbone, which in birds is indicative of having the ability to fly, or in the case of flightless birds is reduced and reflects the loss of their ability to fly.
Terrestrial - Living on the land, ground-dwelling.
Theropods - Carnivorous dinosaurs unrelated to the avian lineage of birds.

ILLUSTRATION CREDITS

All photos, illustrations, sculptures and charts are by Stephen A. Czerkas unless noted otherwise below. Page **1**: *Ornitholestes* by Charles R. Knight, circa 1903, Courtesy of The American Museum of Natural History. **2**: Top left, After Johann Jakob Scheuchzer in Physica Sacra, 1728; Bottom, After John Martin, circa 1844. **3**: Top and middle by B. Waterhouse Hawkins, circa 1854; Bottom, Anonymous. **4**: *Compsognathus* skeleton after Anonymous in John H. Ostrom, 1978, in The Osteology of *Compsognathus longipes* Wagner, Zitteliana 4. **5**: Top, *Archaeopteryx* after Richard Owen, 1863, in On The *Archaeopteryx* of von Meyer, Philosophical Transactions, VOL. 153, Part 1; Bottom, *Archaeopteryx* after Gerhard Heilmann,1927, in The Origin of Birds, D. Appleton and Company. **8**: Top, ibid.; bottom, *Deinonychus*, after Robert Bakker, 1969, in Osteology of *Deinonychus antirrhopus* by John H. Ostrom, Bulletin 30, Peabody Museum of Natural History.